T0065286

AN INTRODUCTION TO AERONAUTICAL STRUCTURES FOR MANAGERS

Som R. Soni

Associate Professor of Systems Engineering (Ret)
Department of Systems and Engineering Management
Air Force Institute of Technology
Wright Patterson Air Force Base, Ohio, 45433, USA

authorHOUSE®

AuthorHouse™
1663 Liberty Drive
Bloomington, IN 47403
www.authorhouse.com
Phone: 1 (800) 839-8640

Published by AuthorHouse 01/14/2016

ISBN: 978-1-5049-6068-7 (sc)
ISBN: 978-1-5049-6069-4 (e)

Library of Congress Control Number: 2015918932

This work is dedicated to my parents
And Samarth Guru
Shri Parthasarthi Rajagopalachari Ji Maharaj.

FOREWORD

Dr. Som Soni has produced a text that would be very valuable to the nontechnical individual who has responsibility in the field of aero structural engineering as related to an overall appreciation of specific topics. He has made the writing very understandable. The topics range from load and stress to composites. In addition, I believe the information discussed would be very apropos for an experienced engineer that wants to review some of the topic areas. I am definitely amazed at how much material is presented by Dr. Soni in a very interesting fashion. He does not skip the more important issues in a particular topic.

Anthony N. Palazotto
Distinguished Professor of Aerospace Engineering
Air Force Institute of Technology
WPAFB, Ohio USA 45433

PREFACE

The material presented in this book was developed for systems engineering and management students in the Air Force Institute of Technology (AFIT) at Wright Patterson Air Force Base (WPAFB), Ohio, USA. This was a one quarter course, MECH 505 entitled "Materials and Structures" in which AFIT report # AF 33(608)-642 authored by H. H. Hurt Jr. was modified and used. Students in the Department of Systems and Engineering Management took course MECH 505 as a preparatory course for structural health monitoring and cost accounting/ composite aircraft cost estimating theses.

This is an assemblage of material for managers, safety officers, contracting officers, accountants, and other related technical and non-technical personnel. It is designed to provide the student with a basic understanding of aero-structures, emphasizing fundamental principles, terminology, clues and factors in effective execution of their duties. Additionally it provides the reader with an understanding of the materials and structures terminology commonly used in the aerospace industry and thus serves as a learning tool for concerned officers preparing them to more effectively and efficiently deal with aircraft maintenance, safety, repair and other relevant management challenges. It is expected that these skills will continue to become more and more useful in the currently exploding requirements within the space and aircraft industry.

We start with definitions of commonly used terms, simple analysis techniques, major causes of failure and relevant clues to understand them and ways to avoid loss of important information. This one quarter graduate level course also prepares individuals for further relevant research work. The table of contents provides an overview of the topics addressed in this text book.

Each year students with a variety of experience and skill sets have contributed to the enrichment of the material and contributed to Master and Doctoral level research. Consequently various theses, dissertations and journal publications have emerged. An impressive list of these is given in Chapter 11. This chapter speaks volumes about the importance of the valuable research being conducted at the university level and its useful nature to the future of the aerospace industry.

Dayton, Ohio   Som R. Soni

ACKNOWLEDGEMENT

This book is a culmination of many years of effort by various individuals including H. H. Hurst, Javier Rodriguez, Matthew Haouser, T. J. Badiru and many of my former students at the Air Force Institute of Technology (AFIT), Wright Patterson Air Force Base (WPAFB) Ohio, USA. I am grateful to each of them for their contributions to this collection of knowledge.

Prior to expanding into this area of mechanics, my research and publications primarily related to mechanics of composite materials and structures. I have been most fortunate to have worked in association with many outstanding professionals in the field of materials and structures for aerospace including the following:

1. Dr. Chandrika Prasad, University of Roorkee, UP, India
2. Dr. Raj Krishan Jain, University of Roorkee, UP, India
3. Dr. C. L. Amba-Rao, Vikram Sarabhai Space Center, Indian Space Research Organization, Trivandrum, India
4. Dr. Jeffrey Warburton, University of Nottingham, Nottingham, UK
5. Dr. Steven W. Tsai, Air Force Materials Laboratory, Wright Patterson FB, Ohio, USA
6. Dr. Nicholas J. Pagano, Air Force Materials Laboratory, WPAFB, Ohio, USA
7. Dr. James M. Whitney, AFML, WPAFB, Ohio, USA
8. Dr. Robert Calico, Air Force Institute of Technology, WPAFB, Ohio, USA
9. Dr. Adedeji Badiru, AFIT, WPAFB, Ohio, USA
10. Dr. Anthony Palazotto, AFIT, WPAFB, Ohio, USA

I am grateful to each of them for providing me the opportunity to work and learn both with them and from them. Thanks also go out to the AFIT faculty, staff and students for all their support during my teaching period there relevant to this book and other courses. Additionally, a special thanks to Dr. Anthony Palazotto for writing the Foreword and Mr. Anthony Hand for editing the final draft.

Further, I take this opportunity to express my thanks to my beloved wife for her patience and perseverance during my extended hours work at home and office.

Dayton, Ohio   Som R. Soni

TABLE OF CONTENTS

1

INTRODUCTION

This chapter provides salient features of aeronautical structures; flight related basic requirements, maintenance and service life consideration as well as definitions of commonly used terms; and lays down the foundation for addressing specific problem areas in the following chapters. The document is prepared to provide practical aspects of theory and application pertaining to the operational problems of air and space vehicles. Only the minimum necessary mathematical relations are provided. For more extensive and complete detail of the advanced phases of structures, materials, and metallurgy the reader is encouraged to use relevant references. This area of study has received lot of attention, therefore a vast collection of books and papers are available. Managers cannot afford to spend too much time to understand detailed treatises of relevant topics and therefore, we consider the following three fundamental objectives in this text:

1. Basic properties of structural materials and their suitability in particular structures.
2. Fundamental reasons for operating strength limitations and good maintenance practices.
3. Provide adequate clues to recognize and diagnose the causes of structural and mechanical failures.

These objectives are first served by describing the principal requirements of any air and space structure. The most important basic requirement is that the primary structure should be the lowest possible weight. All of the basic items of performance and efficiency of a configuration are seriously affected by the structural weight. This is especially true when the extremes of performance are demanded of a configuration. For example, during preliminary design of a long range jet aircraft, a configuration weight growth factor of twenty may be typical. In other words, if the weight of any single item (e.g., landing gear structure) were to increase one pound, the gross weight of the aircraft must increase twenty pounds to maintain the same performance. Any additional weight would require more fuel, more thrust, larger engines, greater wing area, larger landing gear, heavier structure, etc. until the aircraft gross weight had increased twenty times the original weight change.

EXTRA STRUCTURE ◫▭▭▭▷ BIGGER ENGINE

BIGGER ENGINE ◫▭▭▭▷ MORE STRUCTURE

MORE STRUCTURE ◫▭▭▭▷ MORE FUEL

Long range missiles and spacecraft usually encounter a design growth factor which is considerably in excess of any typical aircraft. Some typical long-range ballistic missiles have demonstrated preliminary design growth factors on the order of eighty to two hundred. Of course, such configurations represent an extreme of performance but serve notice of the great significance of structural weight. A limiting situation can exist when demands of performance exceed the state-of-the-art. If performance demands are extreme and basic power plant capabilities are relatively low, the growth factors approach infinite values and impractical gross weights result for the configuration.

The primary structure must be the minimum weight structure which can safely sustain the loads typical of operation. The actual nature of the most critical loads will depend, to a great extent, upon the design mission of the vehicle. During design and development, the mission must be thoroughly analyzed to define the most critical loads which will determine the minimum necessary size and weight of the structural elements. From an apparent infinite number of possible situations, the most critical conditions must be defined. Usually aircraft companies have established cost factors depending upon the maximum speed and weight requirements. Generally, there are three important areas of structural design: (1) static strength (2) rigidity and stiffness (3) service life considerations. Any one of these elements or a combination of them could provide the most critical requirements of the structure. These aspects make this text very useful for managers involved in related fields.

1.1 STATIC STRENGTH CONSIDERATIONS

Static loads refer to those loads which are gradually applied to the structure. The effects of the onset of loading or the repetition of loading deserve separate consideration. Throughout the operation of its mission, a vehicle structure encounters loads of all sorts and all different magnitudes. Various loads may originate during manufacture, transport, erection, launch, flight gusts, maneuvers, landing, etc. These various conditions may be encountered at various gross weights (e.g., positions, altitudes, pressurization, etc.). If particular elements of the structure are separated for study, it is appreciated that these elements are subject to a great spectrum of varying loads.

For the considerations of static strength, it is important that this spectrum be analyzed to select the maximum of all loads encountered during normal intended operation. This maximum of all normal service loads is given special significance by assigning the nomenclature of limit load. The specific requirement of the structure is that it must be able to withstand limit load without ill effect. Most certainly, the structure should not fail at limit load. The primary structure must withstand limit load without undesirable permanent deformation.

Specific requirements are different for various structural applications and in some cases; a yield factor of safety of 1.15 must be incorporated. This requirement would demand that the primary structure be capable of withstanding a load fifteen percent greater than limit without yielding or deforming some objectionable amount. If such requirements were specified for a fighter aircraft, the aircraft could be safely maneuvered to limit "G" without causing the aircraft to be permanently deformed. If such requirements were specified for a typical missile, the missile could be fueled and static tested without causing the structure to be permanently deformed. The number of times this action could be repeated without ill effect would not be part of the static strength consideration.

A separate provision must be made to account for the possibility of a one-time application of some severe load greater than limit. For example, the previously mentioned fighter aircraft may require some flight maneuver load greater than limit in order to avert a disaster of collision. The same idea applies to the missile where malfunction of equipment may cause higher than normal tank pressurization. In either of these examples, some load greater than limit is always a (remote) possibility and, within reasonable limits, should not cause a catastrophic failure of the primary structure. There must be some provision for the rare possibility of a single critical load greater than limit.

Experience with piloted aircraft has shown that an ultimate factor of safety of 1.5 is satisfactory. Thus, a primary structural element should be capable of withstanding one load fifty percent greater than limit without failure. Of course, loads which generate stresses greater than the yield point will cause objectionable permanent deformation of the structure and render it unsuitable for continued operation. The principal concern is that the primary structures withstand the ultimate load without failure. To be sure, the ultimate load can be resisted only by a sound structure (i.e., no cracks, corrosion, eccentricity, etc.).

In the previous discussion, the yield factor of safety of 1.15 and ultimate factor of safety of 1.5 have been selected since these values are representative of piloted aircraft. On the other hand, certain missile configurations may have factors of safety well below that of piloted aircraft (e.g., yield factor of safety of 1.0 and ultimate factor of safety of 1.2). In order to complete the picture, some ground support equipment may have an ultimate factor of safety of five or six and a typical bridge structure may have an ultimate factor of safety of twenty. The factors of safety for airborne vehicles must be as low as is consistent with the safety and integrity of the structure.

When specific limit loads yield and ultimate factors of safety are defined, there will be no deliberate addition of strength above these specified minimums. The reason is simple as undesirable structural weight would be added. However, the normal variation of material strength properties must be accounted for by designing to minimum guaranteed strength or specific levels of probability. In either case, it is possible that a considerable percentage of the structures will exceed the strength requirements by slight margins. This is an expected result when the structure is required to meet or exceed the minimum specified values.

As a result of these static strength considerations, the primary structure must withstand limit load without objectionable permanent deformation and ultimate load without failure. Because of the yield factor of safety and certain material characteristics, objectionable permanent deformation does not necessarily take place immediately above limit load. This could lead to difficulty in appreciating overstress conditions since objectionable permanent deformation does not necessarily occur just beyond limit load. However, at ultimate load, failure is imminent.

1.2 RIGIDITY AND STIFFNESS CONSIDERATIONS

Strength could be defined as the resistance to applied loads. On the other hand, stiffness could be defined as the resistance to applied deflections. This particular distinction between strength and stiffness is a necessary consideration since the development of adequate strength does not insure the attainment of adequate stiffness and rigidity. In fact, the particular requirements of stiffness must be given consideration which is separate (but not completely independent) from the basic strength considerations.

The stiffness characteristics of a structure are very important in defining the response of the structure to dynamic loads. In order to distinguish a dynamic load from the ordinary static load, inspect Figure 1.1 where a weight W is suspended by a spring which has a stiffness k. If the weight was lowered slowly and the force gradually applied to the spring, the spring would deflect slowly until the spring supports the entire weight with equilibrium deflection δ.

$$\delta = \frac{w}{k}$$

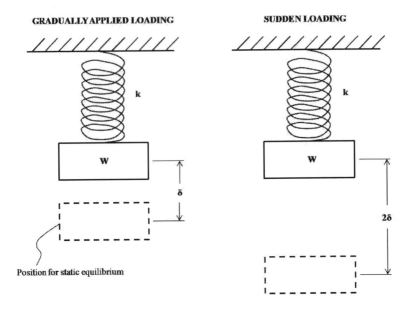

GRADUALLY APPLIED LOADING   SUDDEN LOADING

Position for static equilibrium

Figure 1.1 Deflection of a spring under gradual
and sudden loading conditions

Alternatively, if the weight is suddenly dropped onto the spring, the input energy of the sudden loading will cause the spring to deflect twice as greatly, 2δ. Then, the weight will oscillate back and forth, finally coming to rest at the same equilibrium deflection as for the gradually applied load. During the first plunge when the spring is deflected 2δ, the spring is subject to an instantaneous load which is twice the weight. Under the dynamic loading illustrated, the dynamic load is twice as great as the static load. Of course, if the weight had been projected downward onto the spring with an initial velocity, the input energy would be greater and the dynamic amplification of load would be increased considerably.

The previous example serves only to point out the serious nature of dynamic loads. In a more typical (and complex) aircraft or missile structure, many degrees of freedom exist and the response may show the coupling between various possible modes of oscillation. In any case, the energy of load input, the rate of onset, and the characteristic response of the structure must be examined to determine the critical amplification of loads. The stiffness characteristics and the existence of damping (either natural or synthetic) must be tailored, if possible, to minimize critical amplification of loads.

Vibration of structures may provide critical situations and create a source of damaging loads. The fundamental nature of a vibrating system is best illustrated by the simple spring-mass system of Figure 1.2. If a weight W is suspended on a spring of stiffness k, a small disturbance of the system will cause the weight to oscillate at some frequency which is particular to spring stiffness and weight.

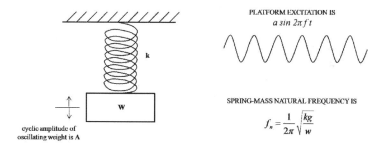

Figure 1.2 Fundamentals of a vibrating system

This natural frequency is related by the following equation:

$$f_n = \frac{1}{2\pi} \sqrt{\frac{kg}{w}}$$

Where f_n = natural frequency
         k = spring stiffness
         g = acceleration due to gravity
         W = weight

In order to consider the possibility of a forced vibration of this system, suppose the platform is moved back and forth with a sinusoidal motion of $a \sin 2\pi ft$, where f is the frequency of excitation and a is the amplitude of excitation. The cyclic motion of the platform will induce a cyclic motion of suspended weight. The amplitude A of motion of the weight is related to the platform excitation and the natural frequency of the spring--mass system. This relationship is defined by the following equation:

$$A = \frac{a}{1 - \left(\dfrac{f}{f_n}\right)^2}$$

Where A = amplitude of weight motion
              a = amplitude of platform motion
              f = excitation frequency of platform
              f_n = natural frequency of spring-mass system

A careful inspection of this equation points out one of the undesirable possibilities of a forced vibration of a structure. As the excitation frequency of the platform equals the natural frequency of the system $(f/f_n) = 1$, a resonant condition develops and the amplitude of weight motion approaches an infinite value. Of course, this resonant condition could cause sudden failure of the spring.

If the platform is subjected to an excitation frequency which is well below the natural frequency, the weight amplitude is very nearly the same as the platform amplitude. The weight would move along with the platform with only slightly greater than static deflection of the spring. When the excitation frequency is considerably greater than the natural frequency, the weight is essentially isolated while the platform oscillates. The cyclic deflection of the spring would approach the cyclic displacement of the oscillating platform. If damping or resistance to motion is introduced into the system, the resonant condition will simply produce less than infinite motion of the oscillating weight. Above and below the resonant condition, damping will alter the motion, depending on the amount of damping present in the system.

While the simple system illustrated does not portray the behavior of complex structures, the fundamental relationships are the same. During design of a structure, the stiffness and mass distribution must be tailored to insure that the ordinary environment of vibration does not allow any approach to resonant conditions and produce damaging loads.

Aeroelastic problems are encountered due to the interaction of aerodynamic forces and elastic deflection of the structure. Since elastic deflections are involved, the inherent stiffness and rigidity of the structure is a principal quality determining the extent of aeroelastic problems.

Static aeroelastic problems involve only the relationship of the aerodynamic forces and elastic deflections without the generation of inertial forces. A typical static aeroelastic problem encountered in aircraft is the phenomenon of *aileron reversal*. Deflection of an aileron produces a section pitching moment tending to twist the wing in torsion. Thus, if an aileron is deflected down at high speed, the wing may develop such significant twisting deflection that the aircraft may roll opposite of the direction desired. Of course, sufficient stiffness must be provided in the structure to prevent aileron reversal or any significant loss of control effectiveness within the intended range of flight speeds. A more disastrous sort of aeroelastic problem is referred to as *divergence*. Suppose that a surface is subjected to a slight up gust when at very high speed. If the change in lift occurs forward of the elastic center, the surface will tend to twist the leading edge up as well as bend it up. The twist represents additional angle of attack (more lift, more bend, more twist, etc.) until a sudden failure results. Such a failure is sudden and catastrophic without warning. It is obvious that divergence could not be tolerated and sufficient stiffness must be present to prevent divergence within the anticipated flight range. In addition, below the divergence speed, there must be sufficient stiffness to insure no serious change in load distribution due to this sort of interaction between aerodynamic forces and elastic deflections.

All of the static aeroelastic problems are specific to the stiffness qualities of the structure and the dynamic pressure of flight (q). Thus, any specific operating limitation imposed will be relative to a certain dynamic pressure (indicated, calibrated, or equivalent airspeed).

The dynamic or oscillatory aeroelastic problems introduce an additional variable, inertial force. Thus, dynamic aeroelastic problems involve some combination of aerodynamic forces, elastic deflections and inertial forces. *Flutter* is one such problem. If a surface with particular mass and stiffness distribution were exposed to an airstream, the oscillatory aerodynamic forces may combine with the various natural oscillatory modes of the surface to produce an unstable motion. Flutter is essentially an aerodynamically excited oscillation in which airstream energy is extracted to amplify the energy of the structural oscillation.

Flutter is not necessarily limited to control surfaces or wing surfaces. Structural panels may encounter flutter conditions which are just as critical and damaging.

During design, the review and analysis of possible flutter behavior constitutes one of the most highly complex studies. The mass and stiffness distribution must be arranged to prevent flutter from occurring during normal operation. There is the implication that any alteration of stiffness or mass distribution due to service operation could cause a possible dangerous reduction of the speed at which flutter would occur.

This fact is important with respect to all of the conditions requiring adequate stiffness of the primary structure. Any alteration of stiffness may produce a dangerous change in dynamic response vibration or aeroelastic behavior.

1.3 SERVICE LIFE CONSIDERATIONS

When considering the service life of a structure, the entire gamut of loads must be taken into account. To achieve satisfactory performance during service operation, a structure must withstand the cumulative effect of all varieties of load that are typical of normal use. Normal periods of overhaul, inspection, and maintenance must be anticipated.

Creep is the continued plastic straining of a part subjected to stress. Of course, creep is of a particularly serious nature when the part is subjected to stress at high temperature since elevated temperatures reduce the resistance to plastic flow. Gas turbine components, reentry configurations, rocket combustion chambers, and nozzles represent some of the typical structures in which creep is important.

If a part is exposed to stress at high temperature, the part will continue to strain at a constant stress. If the exposure time is increased to some critical point, the creep rate will suddenly increase and failure will occur. Such failures due to creep can take place well below the static ultimate strength of the material. Of course, the creep stress must be supplied for sufficient time to generate the condition of failure. It is typical of all metals that any increase in applied stress or temperature will increase the creep rate and reduce the time required to cause failure.

The creep damage is accumulated throughout the life of a structure with the times at high stress and temperature causing the most rapid rate of accumulation. In order for a structure to perform satisfactorily in service, the spectrum of varying loads and temperatures must not cause a critical accumulation of creep damage. In other words, service use should not cause either creep deformation sufficient to prevent operation or creep failure by fracture or buckling. In some applications of turbine machinery and mechanisms, the limit of creep deformation may be the appropriate design consideration. The primary airframe structure may not be adversely affected by such creep deformations and the final failure by fracture or buckling may be the critical consideration.

In special high temperature structures, the anticipated service life has distinct limitations. For example, gas turbine power plants may have turbine structural elements which must be replaced at regular intervals while the shaft case compressor withstands only ordinary inspection. If certain turbine elements were required to demonstrate the same time life as less highly stressed or heated parts, such life may not be at all possible. As a result, the design service life of such high temperature, high stress parts may be set by design limitations rather than arbitrary desirable values.

Fatigue is the result of repeated or cyclic loads. If a metal part is subjected to a cyclic stress of sufficient magnitude, a crack will eventually form and propagate into the cross section. When the remaining cross-section cannot withstand the existing loads, a final rupture occurs as if by static load. The most important aspect of fatigue is that the failure is progressive by the accumulation of fatigue damage. When a critical level of damage is accumulated, a crack forms and propagates until final failure takes place. When a part is exposed to the variety of repeated loads during service operation, the cumulative fatigue must be limited so that failure does not occur within the anticipated service life.

In order to prevent fatigue failures, the structural design must bring into consideration many important factors. First, a reasonable estimate of the service life must be made and the typical spectrum of service loads must be defined. Then, the fatigue characteristics of the materials must be determined by laboratory test of specimens. The effect of stress concentrations, corrosion, environment, residual stresses, and manufacturing quality control must be analyzed. With these factors known, the concept of cumulative damage can be applied to determine the dimensions of the part necessary to prevent fatigue failure during the anticipated service life.

The normal scatter and variation in material properties encountered in fatigue tests will not allow prediction of the specific life of individual parts. A more appropriate consideration would be to account for the variability of material characteristics and load spectrum by the definition of failure probability. In other words, as parts in service approach the design service life, the probability of fatigue failure increases. If parts are exposed to service well beyond the design service life, failure probability will be quite high and the incidence of failures and malfunction of equipment will increase.

The use of periodic inspection and maintenance is necessary to insure failure free operation. Regular inspection must guarantee that parts do not incur excessive deformation or cracks during exposure to fatigue and creep conditions. There is always the possibility of short periods of high stress or temperature which could cause acceleration of creep or fatigue damage and precipitate a premature failure. This is a very important obligation of the maintenance facility.

The various considerations of static strength, stiffness and service life will all contribute specific demands on a structure. Just which of these considerations predominate will depend upon the exact nature of the structure. An aircraft may show that any one of the static strength, service life or stiffness requirements may predominate. In the design of most power plant systems, the service life considerations of creep and fatigue usually predominate. Very short life missile structures may show that the static strength considerations prevail during design. However, if the missile must withstand considerable transportation handling and continuous functional checks, service life considerations may be important.

1.4 COMMONLY USED TERMS AND DEFINITIONS

The function of a manager or investigating officer is to assist his commanding officer in detecting and evaluating hazards, managing safety information and managing risks. This text provides with tools to assist him in following tasks:

Discuss aero-structural problems with engineers, using the terminology of the discipline intelligently.

Read, understand, and interpret to his command, technical directives and messages, which relate to aero-structures.

Conduct a preliminary evaluation of evidence following an aircraft mishap, screen failed parts for cause of failure, and where necessary request assistance from (and supply proper information to) appropriate authorities. During evidence gathering, make sure not to touch fractured surface. This evidence gathering and preliminary evaluation does not make the reader a failure analyst. The failure analyst will conduct a more in depth evaluation utilizing an electron microscope and other scientific tools.

The investigator needs to recognize anomalies, preserve the fracture surface, and present it unaltered to the Failure Analyst for his examination. In particular, never put pieces together.

1.4.1 Fundamental Requirements

Four Fundamental Requirements of an Aircraft Structure are: 1) adequate strength, 2) adequate stiffness, 3) adequate stability, 4) minimum weight. These terms are defined below.

1) **Adequate strength:** Component should not break under design load. Strength is equal to its ability to withstand stress (to be defined). Note that we will define later the term called, "Design Load".

2) **Adequate stiffness**: Component can't deform unacceptably. For example, when the wall is pushed, it pushes back. It deforms but within acceptable limits.

3) **Adequate stability**: Component can't buckle unacceptably. Structure stability means the structure does not buckle. "Structural Instability" implies buckling of the structure. We will learn in later chapter how "buckling" differs from "deflecting".

4) **Minimum weight**: Weight consciousness is a key characteristic of Aero-structural Engineers - if their design weighs too much, it won't fly!

1.4.2 Definitions of commonly used terms

(a) Aerospace structure
(b) Load
(c) Axial load
(d) Shear load (direct shear load)
(e) Torsion, or Torque
(f) Bending moment
(g) Internal force
(h) Concentrated load
(i) Distributed load

(j) Limit Load
(k) Design Load
(l) Factor of Safety
(m) Margin of Safety
(n) Span Loader
(o) Growth Factor

Aerospace Structure: According to the Oxford Dictionary, "Structure is a fabric or framework of material parts put together." Aircraft engineering definition of structure is those parts of an aircraft, the primary purpose of which is to insure the integrity of the aircraft and to carry (i.e., to transfer) the loads encountered in flight and on the ground. Is wing skin a structure? Yes, wing skin is a structure.

Loads: represent the forces, which are applied to the structure of an aircraft, due to gravity, aerodynamics, and inertia. Types of Loads include axial, transverse, shear, bending, torsional, concentrated and internal.

Axial load: is aligned with axis of the structural element as given in figure 1.3. Tension is considered positive, and compression is negative. A bar in figure 1.3 is carrying one P of axial load.

"Direct shear" load, across axis causes surfaces to slide across each other as they come apart – SHEAR! (i.e. figure 1.4)

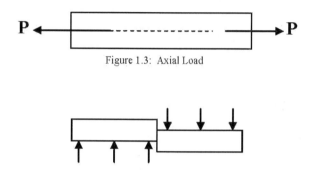

P ←⎯⎯⎯⎯⎯⎯⎯⎯⎯⎯⎯⎯⎯→ P

Figure 1.3: Axial Load

Figure 1.4: Shear Load

Torsion-twisting moment: is shown in figure 1.5. The twisting moment is defined by the formula T=FxL, where F is the force and L is the length shown in figure 1.6.

Figure 1.5: Torque Load

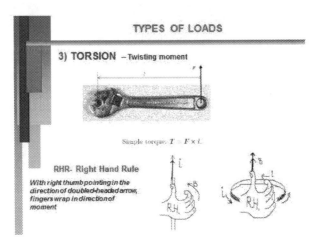

Figure 1.6: Torsion load and corresponding rules.

Right hand rule is defined by using right thumb pointing in the direction of double-headed arrow, fingers wrap in direction of moment.

Bending moment: In case of concentrated load at the tip, shear stress is constant and bending moment increases as you move from tip towards root. In case of distributed load, shear increases as you move from tip towards root. Both arm & resultant are increasing. Thus bending moment takes over the problem.

An internal force is the force carried by a structural member, i.e. the force transmitted by the member. Internal forces are evaluated using equilibrium principles. First, make an imaginary cut at the point of interest and then replace the part that you cut away with an internal force, which will restore equilibrium. Internal forces are important because they allow one to calculate stresses, and thus determine the amount of material required.

Load used in aircraft design: World is too complicated to analyze therefore we analyze Mathematical models of the world.

Concentrated: load is an idealization of load distributed over a small area. It is used because it makes the math easy. **"Load Resultant"** is a fictitious concentrated load which has the same net effect as a distributed load, which it represents. These simplifications are used to make the mathematics easy.

Distributed: load is one spread over a broad area and is an inherent part of lift. We must model the distribution of load.

Limit Load: is maximum load "expected" in service and is equal to the number of g's (representing aircraft speed limit) times the weight of the aircraft.

Design load: is the product of limit load and Factor of Safety. We were not referring to this "design load" definition when we were talking about "adequate strength".

Factor of Safety: is a multiplier used to scale up the limit **load** (or **the** limit stress) for comparison with the material's ultimate load (or ultimate stress) to allow for unknowns in design, manufacture, and mission change. For manned aircraft the factor of safety is 1.5 and that for missiles is 1.25. Figure 1.7 gives a graphical approach to load description.

Yield Factor of safety: is a multiplier used to scale up the limit load (or the limit stress) for comparison with the material's yield load (or yield stress) to insure that unacceptable permanent distortion of the structure will not be encountered. For manned aircraft the factor of safety is 1.15 and that for missiles it is 1 (one). We won't discuss yield factor of Safety now because we haven't defined "yield" yet, we will mention it again later, in chapter 5. Don't confuse factor of safety with margin of safety which is different as defined below.

Margin of Safety: Ultimate Load (ULT) minus Design Load (DEL) divided by Design Load. Margin of safety is usually expressed as a percentage. Margin of Safety is part of the design process; the smaller the better, usually not negative. This affects performance, weight, and cost.

$$\text{MARGIN OF SAFETY} = \frac{ULT - DES}{DES} \times 100\%$$

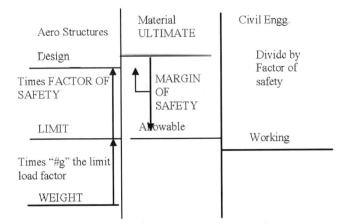

Figure 1.7: Graphical approach to Load Descriptions.

Span Load: Effect of moving loads from fuselage to wings is shown in figure 1.8 which Includes "Span loader concept."

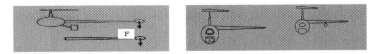

Figure 1.8: Bombs in fuselage vs. bombs on
wing stations (identical aircraft).

Growth Factors: Additional structures require bigger engine, which requires more fuels, which requires more structure. Growth factors are at least 50 to 1 for aircraft; and between 50 and 100 to 1 for spacecraft

2

LOAD AND STRESS DISTRIBUTION

Fundamentally, there are two types of loads which can be applied to an element of structure. There are *axial loads* (often referred to as normal loads) which are applied along the axis of the part, and there are *transverse loads* (often referred to as shear loads) which are applied normal or perpendicular to the axis of the part. Any complicated load condition to which a structure is subjected can be resolved to the various axial or transverse loads acting on a particular part. Since the inherent strength properties of a structural material are based upon the element strengths of its crystals and grains, a more appropriate definition of a load condition on a part is the amount of load per unit of cross-sectional area. Thus, load per unit area is referred to as *stress* and all basic properties of structural materials are based upon stress. Of course, as there are two basic types of loads: axial and transverse, there are two basic types of stress which result from these loads: axial (or normal) stresses of tension and compression and the transverse (or shear) stresses. Axial tensile loads applied to a material will produce certain types of effects and certain types of failures, while axial compressive loads in the same type of material will produce completely different effects and different types of failures. Tensile and compressive axial loads are shown in Figure 2.1.

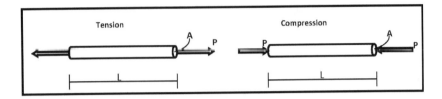

Figure 2.1 Axial loads in tension and compression

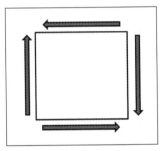

Figure 2.2 Shear loads on an element

A shear stress applied to a particular material result in a somewhat unusual pattern. As depicted in Figure 2.2, if a shear load is applied to an element of material in a vertical direction, that element will experience balancing shear loads in a horizontal direction of an equal magnitude. This occurs to insure that the element is in equilibrium with respect to rotation.

In any case in which the basic loads, either axial or transverse, are resolved on a cross-section of a part, the stresses are then computed for each element of the part as the amount of load per unit area. Simply, load divided by area is stress.

Normal Stress   Shear Stress

$$\sigma = \frac{P}{A} \qquad\qquad \tau = \frac{P}{A}$$

σ = normal stress [psi] τ = shear stress [psi]

P = normal load [lbs] P = shear load [lbs]

A = cross-sectional area [in²]   A = cross-sectional area in²]

As an example of the typical stress distribution in a loaded structure, the following example problem is provided. This example will best furnish an interpretation of the idea of stress and the function of certain structural components.

Figure 2.3 illustrates a typical beam structure. In this case, there is a simplified spar or beam subjected to a concentrated load at the right end. The left end is mounted or fixed to a rigid wall surface. For simplicity, it is assumed that the spar is the entire effective structure and that all loads will be resisted by this part. The problem will be to investigate the stresses at points A and B in this spar structure which result from the application of the shear load at point C. The beam has a constant cross-section throughout the spar as shown in Figure 2.3. The spar flanges in this type of structure furnish the primary bending resistance, while the spar web connecting the spar flanges provides the primary resistance to shear loads. The rivets attach the spar web to the spar flange and there are vertical stiffeners attached to the web to maintain the form, shape, and stability of the structure. The distribution of stress at point A in the beam is best visualized by taking a section through the beam at point A and supplying the internal loads at section A which are necessary to resist the applied external shear load at C, thus maintaining equilibrium of the structure. This is shown in Figure 2.4.

To maintain equilibrium in a vertical direction, there must be a resisting shear load on section A of 10,000 lbs. in a vertical direction down which resists the applied shear load up at point C. Due to the lever arm of force at point C, there is a bending moment produced in the structure at point A. The magnitude of this bending moment is the product of the force and the lever arm (i.e., 10,000 lbs times 100 inches equals 1,000,000 in-lbs.) of moment at section A.

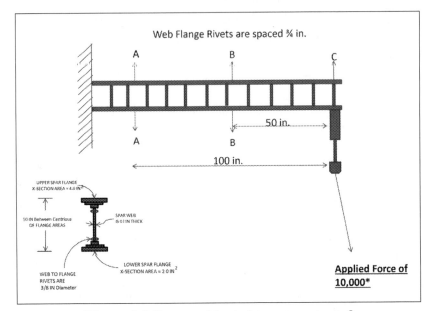

Figure 2.3 Beam subjected to concentrated load with corresponding cross-section.

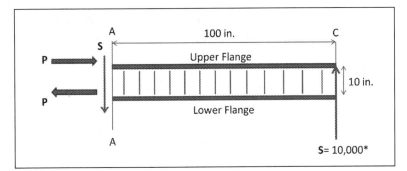

Figure 2.4 Beam sectioned at point A

As the spar flanges provide the primary resistance to bending, there will be a compression axial load developed in the upper spar flange and a tension axial load produced in the lower spar flange. These axial loads in the spar flanges for this untapered beam must be equal to maintain equilibrium of the structure in a horizontal direction. These axial forces in the spar flanges will be referred to as P pounds of load. These two forces of P acting as a couple at a distance of 10 inches apart (the distance between the centers of gravity of areas of the upper and lower spar flanges) must provide internal balance in the structure to the applied external bending moment of 1,000,000 in-lbs. In other words, P multiplied by 10 inches must equal 1,000,000 in-lbs. The P pounds of load in the flange are then computed to be 100,000 lbs.

(P lbs) (10 in) = 1,000,00 in – lbs

P = 100,000 lbs

To determine the stress in the upper flange, the compression load is distributed over the compression area:

$$\sigma_c = \frac{P}{A} = \frac{100,000lb}{4in^2} = 25,000\,psi$$

The tension stress in the lower flange is load divided by area:

$$= \frac{P}{A} = \frac{100,000lb}{in} = 50,000\,psi$$

Since the spar web furnished primary resistance to shear loads, there is a shear load of 10,000 lbs. acting on the effective area of the web. The effective area of this web is the depth times the thickness. In this case, depth is taken as 10 inches and the thickness is 0.10 inches, which produces 1 square inch of cross-sectional area. The shear stress in the web is then load divided by area:

$$\tau = \frac{P}{A} = \frac{10,000lb}{1in^2} = 10,000\,psi$$

To investigate the stress distribution at section B, the same fundamental procedure is employed. The section at point B is furnished in Figure 2.5 with the loads on the cross-section necessary to place the structure in equilibrium. The bending moment to be resisted by the spar flanges is the 10,000 lbs of force acting at the 50 inch lever arm. This produces a bending moment of 500,000 in-lbs and results in axial loads in the spar flanges of 50,000 lbs each.

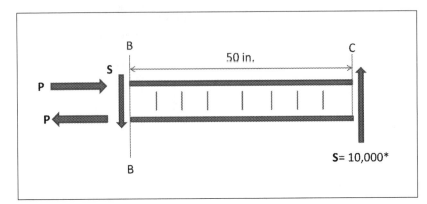

Figure 2.5 Beam sectioned at point B

The compression stress in the upper flange is:

$$\sigma_c = \frac{P}{A} = \frac{50,000lb}{4in^2} = 12,500\,psi$$

The tension stress in the lower flange is:

$$\sigma_i = \frac{P}{A} = \frac{50,000lb}{2in^2} = 25,000\,psi$$

Since the same amount of shear load must be supplied on section B to provide equilibrium in a vertical direction and resistance to the applied 10,000 lbs shear load, the shear stress in the spar web remains the same 10,000 psi throughout the span of the beam and it does so as long as there is no change in the shear load across the beam.

To determine the stress in the rivets attaching the spar web to the flanges, the function of the rivets must be made clear. The point of load application at section C on the beam has a vertical load of 10,000 lbs. applied. This concentrated load of 10,000 lbs. must be appropriately distributed to the web by a fitting. As depicted in Figure 2.6, the desired result is to distribute the concentrated shear load of 10,000 lbs. to the edge of the spar web such that for the 10 inches effective depth of the web, there is 1,000 lbs. of load for each inch along the edge.

Element 1 in the upper right hand edge of the web at section C has applied on its edge a shear load of 1,000 lbs. Figure 2.7 illustrates the manner in which this 1,000 lbs. of load applied to this one-inch element 1 is resisted.

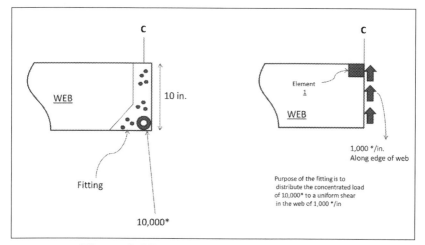

Figure 2.6 Load distributions at section C

At the left hand side, there is a shear load down of 1,000 lbs balancing the applied 1,000 lbs. This shear load on the left hand edge is furnished by the adjacent element of the web to the left of element 1. Since an element of structure with an applied shear load must also have balancing shears at 90°, there will exist (or must exist) on element 1 shear loads on the upper and lower edges as shown.

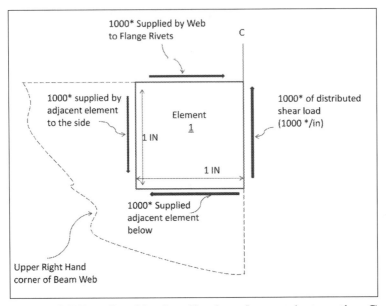

Figure 2.7 Details of loads affecting element 1 at section C

The shear load on the upper edge of element 1 is supplied by the next piece of structure in contact with the edge of the web. This load must come from the spar flange and is transmitted by the web to the flange rivets. The primary function of the web to flange rivets is to provide a continuity of shear and to balance the applied vertical shear load in a horizontal direction. The shear load on the lower side of the element is supplied by the next element immediately underneath element 1 in the web. If the spacing of rivets attaching the web to the flange is 0.75 inches, the load for each rivet in shear will be 0.75 of 1,000 lbs or 750 lbs per rivet. With the diameter of the rivet given as 0.375 inches, the rivet stress in shear is then computed as the load divided by area:

$$Rivet\ Stress = \tau_r = \frac{P}{A} = \frac{750lb}{\frac{\pi}{4}(0.375in)^2} = 6,790.6\,psi$$

Figure 2.8 gives an illustration of the upper and lower spar flange removed from the structure and the loads applied to the spar flanges and the web as illustrated above. The flange loads of 100,000 lbs at section A is the result of the accumulated axial force from the distributed shear loads of 1,000 lbs per inch for the entire length of 100 inches. Thus the distributed load is 100,000 lbs at section A (100 inch length) and 50,000 lbs at section B (50 inch length).

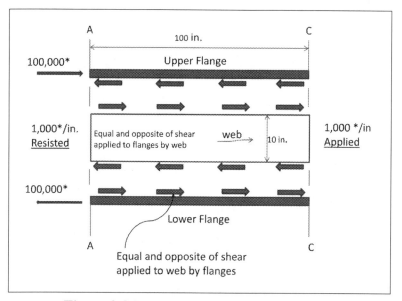

**Figure 2.8 Beam showing loads with upper
and lower spars removed from the web**

The vertical stiffeners (figure 2.9) attached to the web have no particular stress (compression, tension, or shear) until the web of the spar begins to buckle.

The primary function of these vertical stiffeners is to maintain the form and stability and to provide support for the web panel, thereby preventing or delaying the shear buckling of the web. When buckling occurs in the web, the vertical stiffeners must withstand compression loads to prevent collapse of the structure.

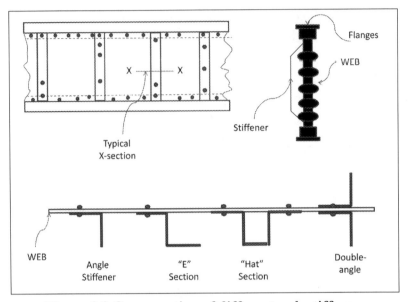

Figure 2.9 Cross-section of different web stiffeners

The function and importance of the spar web can be best emphasized by two examples shown in Figure 2.10. If there were no web between the pinned flanges and a shear load was applied, no resisting loads would be developed in the flanges. The structure would collapse to the shape shown with no resistance to the applied load. If there were no web between the fixed flanges, a shear load applied would produce secondary bending of the flanges. Since the bending resistance of the flanges is quite small, prohibitive stresses and deflections would result. Obviously, the more efficient structure is the web-flange combination which resists shear in the web and bending by axial loads in the flanges.

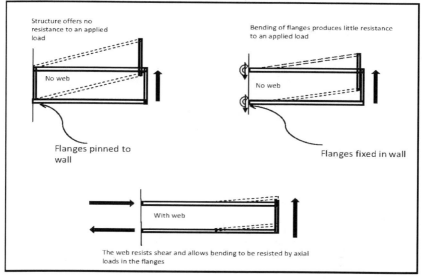

Figure 2.10 Shear load applied to flanges without and with a web

There may be such an object as an access hole in a shear web. These holes are typically used for access in maintenance and production functions and are a structural penalty for anything but minimum gage thickness structures (light planes, gliders, airships, etc.).

The previous example problem of stress distribution in a simplified structure has considered that there were no particular complications to produce anything other than pure axial and shear stresses. In a more detailed analysis, consideration would be made for the contribution of the web to bending resistance and the complication and magnification of stress due to rivet holes, etc.

There are, of course, examples in which the distribution of stress in a structural element is complicated by the particular manner of loading. Certain examples of the particular stress as distributed in typical structures if there is bending, torsion, etc. and the resulting stresses remaining in the elastic range of the material characteristics will be studied next. By elastic range, it is implied that stresses may be applied, and then released, and no permanent deformation of the structure would be incurred.

2.1 PURE BENDING

Figure 2.11 illustrates the case of a solid rectangular bar with pure bending moments applied. The stresses at section A will be distributed as shown with the upper portion of section A subjected to a compression stress which will be a maximum at the outer surface, and the lower surface of section A subjected to a tensile stress which will be a maximum at the outer surface.

The point at which the stress is zero (neither tension nor compression) is referred to as the neutral axis. For a symmetrical, rectangular section, this point would be midway between the upper and lower surfaces. The neutral axis for a homogeneous material subjected to elastic bending is always located at the center of gravity (or cancroids) of cross-sectional area.

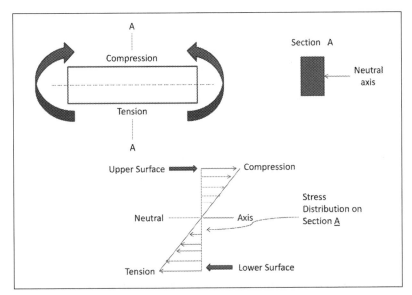

**Figure 2.11 Solid rectangular bar with
pure bending moments applied**

In Figure 2.12, there is a beam with an unsymmetrical cross-section subjected to pure bending. For the cross-section shown, the neutral axis will be closer to the upper surface of the part than the lower surface. The stress distribution illustrated in Figure 2.12 will continue to be a linear variation of stress between the two maximum stresses at the upper and lower surfaces. However, as the neutral axis is closer to the upper surface, the magnitude of compression stress will be smaller than the tensile stress on the lower surface. This must occur since the compression load produced by the smaller compression stress distributed over the larger area above the neutral axis will be equal to the tension load produced by the higher tensile stress acting over the smaller tensile area below the neutral axis. Thus, equal and opposite compression and tensile loads exist which furnish equilibrium to the cross-section in the horizontal direction.

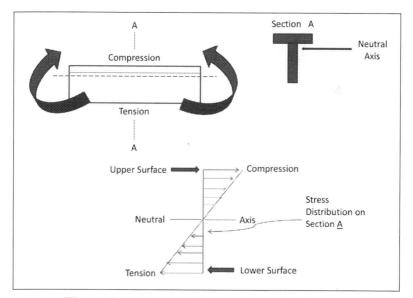

Figure 2.12 Beam with unsymmetrical cross-section subjected to pure bending

The maximum bending stress for beams or other structures subjected to bending moments is calculated as:

$$\sigma_{b\,max} = \frac{Mc}{I}$$

Where
$\sigma_{b\,max}$ = maximum bending stress [psi]
M = bending moment on cross-section [lb-in]
c = distance from neutral axis to outermost fiber [in]
I = moment of inertia of cross-section [in^4]

If the stress on some other fiber is desired, the equation becomes:

$$\sigma_b = \frac{M \, y}{I}$$

Where, y = distance from neutral axis to fiber under consideration.

The term *moment of inertia* appears in the previous equations. As has been explained, the bending stress in a fiber depends on its distance from the neutral axis. If the material can tolerate a given amount of stress, the largest moment can be resisted when the area resisting it is as far as possible from the neutral axis. The moment of inertia is a quantity which takes the shape of the cross-section into account in determining the amount of bending moment which can be resisted by a given cross-section without exceeding a specified stress. For several cross-section shapes, the moments of inertia are:

Rectangular	Triangular	Circular	Doughnut
$I = \dfrac{bh^3}{12}$	$I = \dfrac{bh^3}{36}$	$I = \dfrac{\pi d_0^4}{64}$	$I = \dfrac{\pi(d_0^4 - d_i^4)}{64}$

Where
I = moment of inertia about the neutral axis [in⁴]
b = rectangle or triangle width [in]
h = rectangle or triangle height [in]
d_o = outside diameter [in]
d_i = inside diameter [in]

2.2 PURE TORSION

Figure 2.13 illustrates the condition of pure torsion applied to a solid, circular shaft. With torsion applied to the shaft, the stress produced on section /A is primarily that of shear, a stress which is a maximum at the outer surface and varies linearly to zero at the axis of the part.

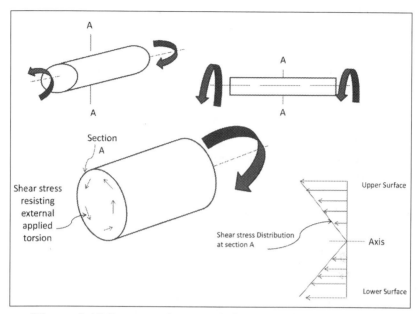

Figure 2.13 Pure torsion applied to a solid circular shaft

For circular shafts subjected to torsion moments, the maximum shearing stress is:

$$\tau_{max} = \frac{Tc}{J}$$

Where τ_{max} = maximum shearing stress [psi]
T = torsional moment or twisting moment [lb.-in]
c = radius from axis to outermost surface [in]
J = polar moment of inertia of cross-section [in^4]

If the stress at a distance r from the axis is desired, the equation is:

$$\tau = \frac{Tr}{J}$$

The polar moment of inertia is used for determining strength of sections subjected to torsion.

Only the circular sections have simple equations for this type of stress distribution. The polar moments of inertia are as follows:

<table>
<tr><td align="center">Circular</td><td align="center">Doughnut</td></tr>
<tr><td align="center">$$J = \frac{\pi}{32} d_0^4$$</td><td align="center">$$J = \frac{\pi}{32}(d_0^4 - d_i^4)$$</td></tr>
</table>

For sections other than circular, reference may be made to any of the more standard texts on strength of materials.

Figure 2.14 illustrates the conditions in which a continuous, hollow cross-section is subjected to torsion. In this instance, the shear stress is a constant value around the periphery of the cross-section. Since the shear load distributed around the periphery must provide an internal resisting moment equal to the external applied moment, the shear stress for this continuous shell structure may be computed by use of the following relationship:

$$\tau_s = \frac{T}{2At}$$

Where
τ_s = shear stress [psi]
T = applied torque [in-lbs.]
A = enclosed area of cross-section [in²]
t = shell thickness [in]

If this section were slotted in a longitudinal direction, a large amount of the torsional rigidity would be lost. There would be no continuous shear stress distributed around the periphery of the shell and high local shear stresses with great deflections would be encountered. The Polar Moment of Inertia, J, is a quantity, which tells you how well the distribution of cross-sectional area carries the torque. The larger the value of J, the better it is. A hollow "torque tube" is the most efficient way to carry torque and the area must be enclosed in order to carry the torque well.

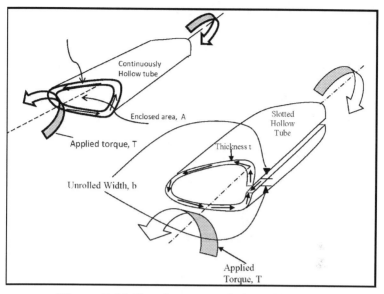

**Figure 2.14 Continuous and slotted hollow
tubes subjected to torsion**

The maximum shear stress in this case could be calculated by application
of the following equation:

$$\tau_{s\,max} = \frac{3T}{bt^2}$$

Where $\tau_{s\,max}$ = maximum shear stress [psi]
T = applied torque [in-lbs.]
b = unrolled width [in]
t = shell thickness [in]

Any shell structure subjected to tensional loading which has a cutout
or slot will have a tendency to develop much higher stresses and may
be excessively flexible. Further, noncircular cross-sections are not used
to carry torque for two reasons: 1) Stress distribution and, 2) warping
out of plane.

2.3 BOLTED OR RIVETED JOINTS

Figure 2.15 illustrates a typical type of bolted or riveted joint which is encountered in more conventional structures. In such a joint, the load applied to element A is transferred by the bolt and distributed to elements B and C. This transfer of load by the bolt produces a shear stress on the bolt cross-section. The bolt shear stress is:

$$\tau_s = \frac{load}{area} = \frac{(P/2)}{(\pi/4)d^2} = \frac{2P}{\pi d^2}$$

The tensile load developed in the plate creates a critical tensile stress along section X. The average tension stress at section X is:

$$\sigma = \frac{load}{area} = \frac{P}{(w-d)t}$$

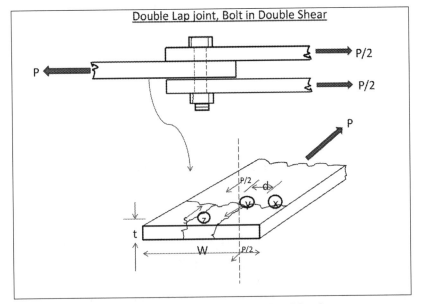

Figure 2.15: Double lap joint, double shear

In addition, the bolt bearing on the surface of the hole (surface Y) may create a critical compressive stress. The average bearing stress in the plate at the bolt hole is:

$$\sigma_b = \frac{load}{area} = \frac{P}{td}$$

The application of the bearing load creates a shear stress along section Z which tends to tear out the edge of the plate. The edge tear out stress is:

$$\sigma_s = \frac{load}{area} = \frac{P}{2ts}$$

These simple equations represent only the simple average stresses in order to appreciate some of the fundamental requirements of a joint. There is the obvious possibility that friction between the plates may accomplish part of the shear transfer and the hole may create considerable stress concentration to cause peak stresses well above the computed average.

2.4 PRESSURE VESSELS

The use of highly pressurized containers in various aircraft and missiles create significant structural problems. Some of the more simplified situations are represented by the stresses created in pressurized spherical and cylindrical shells. These vessels are illustrated in Figure 2.16.

The stress in a pressurized spherical shell is uniform and constant in all directions along the surface. The resulting tensile stress is due to the pressure load (p) being distributed over the effective shell area:

$$\sigma_{spherical\ shell} = \frac{load}{area} = \frac{(pressure)(area)}{area} = \frac{p(\pi r^2)}{2\pi rt} = \frac{pr}{2t}$$

In order to consider the stresses in a pressurized cylindrical shell, the existence of two separate stresses must be noted. The longitudinal stress (σ_x) is related as follows:

$$\sigma_{x\ cylindrical\ shell} = \frac{load}{area} = \frac{(pressure)(area)}{area} = \frac{p(\pi r^2)}{2\pi rt} = \frac{pr}{2t}$$

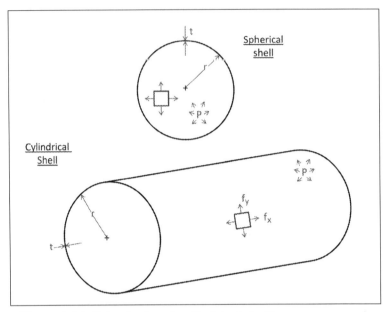

Figure 2.16: Spherical and cylindrical shell pressure vessels

The result obtained is identical to the relationship developed for the pressurized spherical shell. However, an additional stress (σ_y) is developed and is referred to as the hoop stress:

$$\sigma_{y\ cylindrical\ shell} = \frac{load}{area} = \frac{2pr}{2t} = \frac{pr}{t}$$

For this situation, the hoop stress incurred is twice as large as the longitudinal stress developed. An important fact is concluded from this relationship: if a failure due to pressure occurs in a uniform cylindrical shell, the hoop stresses will predominate in the mode of failure.

2.5 SUMMARY:

In this section, the following commonly used terms are defined:

2.5.1 Stress ("Stress Component"): Stress Vector is the force per unit area applied to a structure defined at a point. Stress, perhaps more properly a stress component, is a component of the stress vector at the point of application. There are two and only two kinds of stress, 1) Normal stress and 2) Shear Stress. "Normal Stress", is a stress perpendicular to the surface of the imaginary cut; and "Shear Stress", is a stress parallel to the surface of the imaginary cut (or "tangent to the surface..." or "lying on the surface..." or words to that effect).

Other names for stress can be considered to be "nicknames" rather than true "kinds" of stress — e.g., "bending stress" is the nickname for "Normal Stress Due To Bending." Average stress is the total stress averaged over a surface area. The units of stress are force per unit area.

Tension/Compression: Axial load, normal stress.

Assume a uniform distribution, of stresses over the cross-section with area A.

Formula: $f_t = \dfrac{P}{A'}$

Where, P is load applied perpendicular to area A, as shown in Figure 2.1. Tension is positive, compression is negative.

2.5.1.1 Stress Computation

Average Shear Stress: Direct shear load.

Assume *a* uniform distribution of stress over the cross-section.

Formula: $f_s = \dfrac{S}{nA}$

Watch out for multiple shear situations, n represents the number of interfaces in the joint.

2.5.2 Torsion of Circular Cross-section:

Question: Describe qualitatively the distribution across the cross section of torsional shear stresses, and explain the implications of this distribution.

Answer: Stress distribution is linear, with zero stress at the center, and maximum stress at the outside edge of the circular cross-section.

Torsional shear stress formula: $f_s = \dfrac{Tr}{J}$

The Polar Moment of Inertia, J, is a quantity, which tells you how well the distribution of cross-sectional area carries the torque. The larger the J, the better it is. A hollow "torque tube" is the most efficient way to carry torque, and one must enclose area in order to carry the torque well.

Question: How about using non-circular cross-section with torsion load?

Answer: We don't use non-circular cross-sections to carry significant amounts of torque for two reasons:

Stress distribution is not symmetric about the axis.

There are chances of warping out of plane.

Conclusion: The best way to carry torque is with hollow circular cross-section.

2.5.3 Bending Stress (Normal stress due to bending):

Question: Describe qualitatively the distribution across the cross-section of beam stresses, both normal and shear, and explain the implications of these distributions.

Answer: Bending causes a linear distribution of normal stress in a straight beam. If the beam is curved when unloaded, a more sophisticated theory is required. On bending a straight beam, it bends into a circular arc with stress distribution the same at each point. Neutral axis is in the middle of the cross-section and the magnitude of stress depends upon the distance from neutral axis.

Bending stress formula: $f_b = \dfrac{My}{I}$

The Moment of Inertia, I, is a quantity which tells how well the distribution of cross-sectional area carries the bending moment. The larger the value of I, the better it is. To calculate I use the relevant formula as provided in the text of this chapter.

Distribution of shear stresses over a cross-section of a beam bent by applied shear load must be zero at both the "top" and "bottom" of the cross-section since there is a free surface. But since there is a resultant, shear stresses can't be zero everywhere. Therefore, there must be a maximum somewhere in between!

For initially straight beams with rectangular cross-sections, the shear stress distribution is parabolic with the maximum shear stress at the center of the cross-section, on the neutral axis.

2.5.4 Implications of the distribution of beam stresses:

Normal stresses are greatest in absolute value at the outer surfaces (tension/compression).

Shear stresses are greatest somewhere in the middle.

The problem is driven by bending (i.e. Normal stress) except for the region close to the tip.

Thus there is extra material, in the middle which could be removed and the result is the I-Beam or the Honeycomb Sandwich as given below.

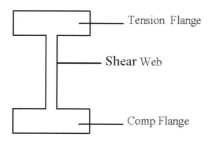

I-Beam   Honcycomb

3

COMPONENT AND PRINCIPAL STRESSES

Two factors determine the strength and manner of failure of any structural member. One is the property and character of the structural material. The other is the maximum normal stresses and maximum shear stresses which exist in various areas of a part. Whenever there is a normal stress applied to a part, there exist various magnitudes of normal and shear stresses at planes different to the direction of loading. Also, any time a shear stress is applied to a part, there exist various magnitudes of normal and shear stresses at certain planes. The various components of the applied primary stress must be investigated to determine the influence upon strength and failure type.

A typical example of component stress is illustrated in Figure 3.1(a) which shows a specimen of material with a pure axial tension load applied. If a section or cut is made in this specimen at section X, it is seen that a constant tension stress exists on this plane perpendicular to the direction of loading. Figure 3.1(b) shows the same specimen subjected to the same pure axial tension load, but with a cut made along section Y. An investigation of the stresses acting on section Y shows that there still exists a uniform tensile stress across the section. However, the tensile stress which exists on plane Y will have components which are perpendicular and parallel to the surface of the cut. One of these components which is perpendicular to the face of the section will be a normal tensile stress of a smaller magnitude than the tensile stress applied to the cross-section. Along the surface of section Y will be a component of stress parallel to the surface. This is a shear stress and it is a component of the primary applied tensile stress.

Since the tensile stress component perpendicular to the face of the cut will be of a smaller magnitude than the applied tensile stress, it will have no particular or immediate influence on strength or mode of failure. However, the shear stress component parallel to the face of the section is a stress of an entirely different nature. It may have a decided effect on strength and failure type as some materials are much more critical in shear than in normal stress, particularly ductile metals.

Figure 3.2(a) shows a part loaded in tension with a section cut along the direction of applied stress. It is obvious there would be no shear stress (due to force components) along section Z. Since shear stresses do not exist either at a section perpendicular to the direction of applied stress or at a section parallel to the direction of applied stress, it is reasonable to assume that between these limits, the existing shear stress will be a maximum on a section at 45° to the direction of primary load application.

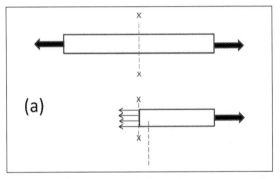

**Figure 3.1(a) Section showing plane perpendicular
to applied axial tensile load**

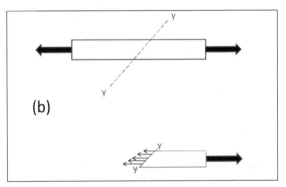

Figure 3.1(b) Section showing plane at 45° to applied tensile load

To determine the magnitude of the maximum shear stress, it is best to take a one inch cube of material from the basic specimen and examine the forces existing in this element. As seen in Figure 3.2(b), there is a load applied to the sides of this one inch element which is the result of the tensile stress on the specimen cross-section. Since the section or cut is to be located at 45° to the applied stress, the component of force (S) distributed along the diagonal will be:

$$S = T \sin(45°) = 0.707T$$

This shear force will be distributed over the diagonal surface, which will be 1.414 inches for the one inch element. The shear stress (τ) is:

$$\tau = \frac{0.707T}{1.414} = 0.5T$$

Thus, an applied axial stress will produce a maximum component shear stress which is half the magnitude of the applied stress. Knowledge of the presence of the component shear stress and its existence as a maximum at 45° is basic to a discrimination between ductile and brittle failure types.

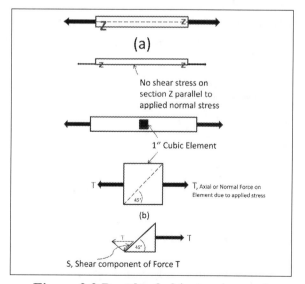

Figure 3.2 Part loaded in tension and element showing component forces

An example of the existence and orientation of the maximum component shear stress is provided in Figure 3.3 where two specimens of metal (one very ductile and one very brittle) are subjected to failing tensile loads. The brittle material will fracture with a clean break at 90° to the direction of loading.

The ductile specimen will exhibit fracture planes at 45° to the direction of the applied tensile stress, thus verifying the existence of component shear stresses. This 45° type of fracture from pure tensile loads is referred to as the ductile or shear type of failure.

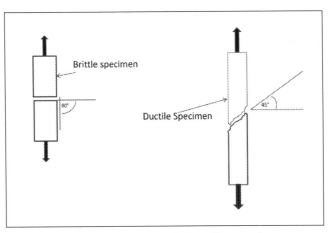

**Figure 3.3: Brittle and ductile specimens
subjected to failing tensile loads**

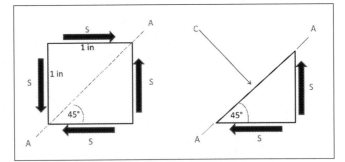

Figure 3.4 Element subjected to pure shear with component forces

The condition of an applied pure shear presents a different and slightly more complex problem concerning component and principal stresses. Consider a one inch cubic element subjected to pure shear as in Figure 3.4.

If a section A is taken at a diagonal of 45° in one direction, it is apparent that there must be a compression force (C) on the diagonal to statically balance the action of the two shear forces applied along the edges of the element. The two shear forces have components at 45° which are additive and must be balanced. The resultant compression force is:

$$C = S\sin(45°) + S\cos(45°) = 1.414S$$

Since the compression force is distributed along the diagonal, the compression stress may be found by taking the force divided by the area:

$$\sigma_c = \frac{1.414S}{1.414} = S$$

For the case of pure shear applied to an element, there exists at one 45° section a compression stress equal in magnitude to the applied shear stress. The existence of this component compression stress is evident in the failure mode of a thin-walled tube subjected to a failing torsion load (torsion produces a uniform pure shear). If the walls of the tube are sufficiently thin, the tube will fail primarily in buckling due to the principal compression stress existing in the 45° direction. This situation is depicted in Figure 3.5.

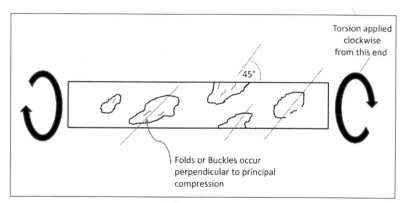

Figure 3.5 Thin-walled tube subjected to torsion load

If the same one inch element of Figure 3.4 is subjected to the same pure shear condition but sectioned along plane B, a different stress situation results. If the element is cut along section B as seen in Figure 3.6, it is apparent that a tension force (T) must be sufficient to balance the components of the two shear forces along the edge. This situation will produce a tension stress distributed along the diagonal which is equal in magnitude to the applied shear stress. The presence of this tension stress is verified by the mode of failure of a brittle shaft in torsion (twisting a piece of chalk produces the same result). As seen in Figure 3.7, a fracture will begin as a 45° spiral surface which is perpendicular to the principal tension stress.

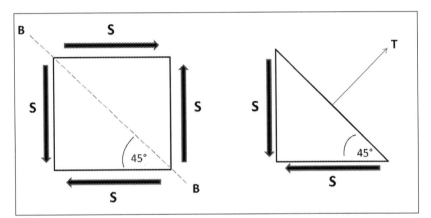

Figure 3.6: Element subjected to pure shear with component forces

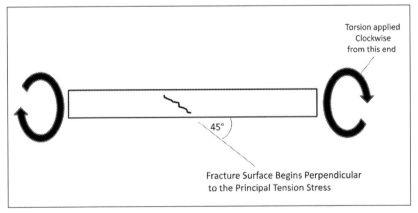

Figure 3.7 Brittle shaft subjected to torsion

It must be remembered that component stresses may have a definite bearing on the strength and mode of failure of any structure. Any normal or shear stress applied to a part will produce component and principal stresses that cannot be neglected.

Summary:

Transformation of Stress:

The type of stress or strain present depends **on the** coordinate system — i.e. on the direction of the imaginary cut. Stress as shown in Figure 3.3 tension bar cut straight across vs cut at 45°. The actual state of stress should be independent of coordinate systems. Thus there must be a way to transform one coordinate system to another. Further, the coordinate system developed is independent of the material system.

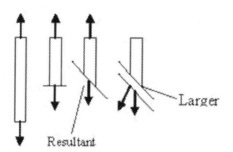

Mathematically, the stress transformation equations are found from equilibrium considerations:

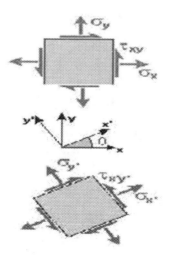

Note that new normal stress receives contributions from old shear stresses as well as old normal stress!

σ_x, σ_y, and $\sigma_{x'}$, $\sigma_{y'}$ are normal stresses.

τ_{xy} and $\tau_{x'y'}$ are shear stresses.

<u>Transformation Equations:</u> A set of second rank tensor transformation equations is given below.

$$\sigma_{x'} = \frac{\sigma_x + \sigma_y}{2} - \frac{\sigma_x - \sigma_y}{2}\cos(2\theta) - \tau_{xy}\sin(2\theta)$$

$$\sigma_{y'} = \frac{\sigma_x + \sigma_y}{2} - \frac{\sigma_x - \sigma_y}{2}\cos(2\theta) - \tau_{xy}\sin(2\theta)$$

$$\tau_{x'y} = -\frac{\sigma_x - \sigma_y}{2}\sin(2\theta) + \tau_{xy}\cos(2\theta)$$

Note: You need not remember anything about these equations, except the fact that they exist!

Important Points to Remember:

<u>Stresses can</u> be <u>transformed</u> from one coordinate system to another.

Shear stress in one coordinate system will contribute to normal stress in a system inclined to it, and vice versa...

YOU CAN'T HIDE; both shear and normal stresses are present when either is present - they're just on different planes!

Fundamental theorem of stress analysis:

"Whatever the state of stress, there will always be three mutually perpendicular planes on which the shear stress components zero, and the normal stress components have stationary values (implies a local maximum or minimum value).

These planes are called PRINCIPAL PLANES, and the normal stresses on them are called PRINCIPAL STRESSES."

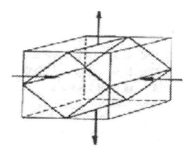

The box in the figure consists of principal planes. The maximum shear planes bisect the angle between the maximum and minimum normal stress planes.

Maximum Shear Stress Corollary:

The Maximum Shear Stress occurs on planes which bisect the angle between the Maximum Normal Stress Plane and the Minimum Normal Stress Plane.

Torsional Failure.

For Brittle Material there is no gross plastic deformation

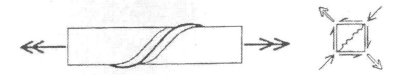

Maximum normal stress plane is oriented 45-degrees to axis of shaft spiral failure. Fracture surface is a Helical – like a screw thread. Surface texture is rough, granulated. The following chalk torsional failure is a good experiment that can be done in the class.

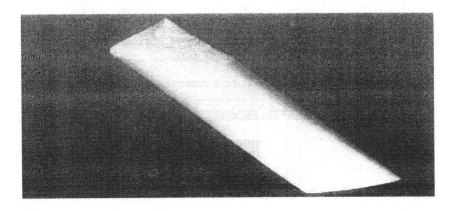

For Ductile Material there is considerable gross plastic deformation, for example twisting of a shaft.

Fracture surface is flat, normal to axis of shaft, and smooth, almost machined looking, shear stress failure.

"Trace marks" are concentric, "tit" in center.

4

STRAIN RESULTING FROM STRESS

Any structure which is subjected to stress must deform under load even though the deformation may not be visible to the naked eye. Recall from the previous sections that stress is the true measure of state for a part subjected to load. Strain is a similar means of measure. In order to fully evaluate the state of a stressed material, all deformations must be considered on a unit basis. Strain is thus defined as deformation per unit length and in most engineering terminology is denoted by ε or e.

Suppose a steel bar 100 inches long is subjected to a tensile stress of 30,000 psi as depicted in Figure 4.1. The total change in length throughout the 100 inches of length would be about 0.1 inches.

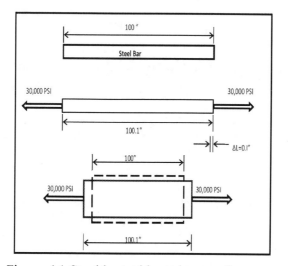

Figure 4.1 Steel bar subjected to tensile stress

If subjected to uniform stresses, the part would be subjected to a uniform strain which is:

$$strain = \varepsilon = \frac{total\ deformation}{original\ length} = \frac{\Delta L}{L} = \frac{0.1 in}{100 in} = 0.001 \frac{in}{in}$$

To produce a total deformation of 0.1 inches in 100 inches of length, the strain must be 0.001 inch/inch or 0.1%.

In addition to the longitudinal strain of the part shown in Figure 4.1, there will be a lateral contraction of the metal. For most metals, there is a definite relationship between this transverse strain and the longitudinal strain and it is important when considering combined stresses that are dependent upon deflections. The proportion between the lateral strain and the longitudinal strain has been given the name Poisson's ratio (denoted by v). This ratio has the approximate magnitude $v = 0.3$ for most homogeneous metals. In the case of the previous example, the longitudinal extension strain of 0.001 would be accompanied by a lateral contraction strain of 0.0003.

It is important to remember that any metal subjected to stress must strain. The amount of strain, while not necessarily visible, must be present and is very important. Just as large stresses may produce numerically small strain, any small strain forced on a structure may produce large stresses.

Shear stresses also produce shear strains. Shear strain, while not necessarily denoting a change in length, does describe a change in relative position of the part. Figure 4.2 illustrates this fact.

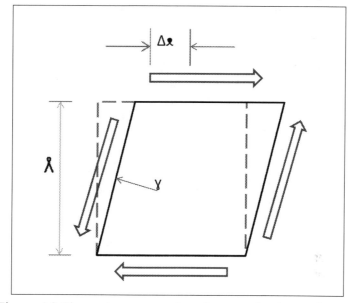

Figure 4.2 Shear stresses causing shear strains in an element

Shear strain is most properly described by the angle of strain (γ) in radians. When a shear stress is applied to an element, the shear strain γ can be computed as the proportion of the change in position of one side (Δ*l*) to the original length (*l*).

$$\gamma = \frac{\Delta l}{l}$$

One of the most important properties of an aircraft material is its stiffness. If a part were subjected to a particular level of stress, small strains would indicate a stiff or rigid material while large strains would indicate a flexible material. The accepted method of measuring the stiffness or rigidity of a material is to compute a proportion between the applied axial stress (σ) and the resulting axial strain (ε). This proportion is known as the Modulus of Elasticity (or Young's Modulus) and is denoted by E.

Modulus of Elasticity = Young's Modulus = $E = \dfrac{\sigma}{\varepsilon}$

Typical values for the Modulus of Elasticity are:

Steel	E = 30,000,000 psi
Aluminum Alloy	E = 10,000,000 psi
Magnesium Alloy	E = 7,000,000 psi

By comparison of these values, it is seen that an aluminum alloy part subjected to a given stress would strain three times more than a steel part subjected to the same stress level.

There is no true modulus of elasticity in shear. However, for computing shear deflections, there exists an equivalent quantity known as the Modulus of Rigidity (or Shear Modulus) denoted by G. The Modulus of Rigidity is the shear stress (τ) divided by the shear strain (γ).

$$Modulus\ of\ Rigidity = Shear\ Modulus = G = \frac{\tau}{\gamma}$$

For most metals, the shear modulus is approximately 40% of the modulus of elasticity:

Steel $\qquad\qquad$ G = 12,000,000 psi

Aluminum Alloy $\qquad$ G = 4,000,000 psi

For homogeneous materials, the Modulus of Rigidity may be determined by the following equation:

$$G = \frac{E}{2(1+v)}$$

Bending stresses will cause bending deflections. In this case, no general strain relationship can be defined that is similar to simple axial strain. Since bending stresses do vary throughout the cross-section, there will be a variation of axial strains proportional to the axial stress.

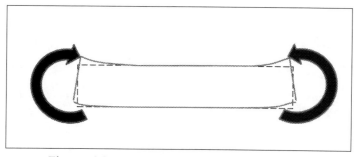

Figure 4.3: Beam subjected to pure bending

For the initially straight beam shown as the dashed line in Figure 4.3, an applied pure bending moment will produce a deflection of pure curvature, denoted by the solid line. Pure bending will cause compressive strains above the neutral axis and tensile strains below it. There will be zero strain on the neutral axis. The distribution of strain is linear: if one end of the beam is held stationary as in Figure 4.4, then the other end deflects upward due to the curvature of bending.

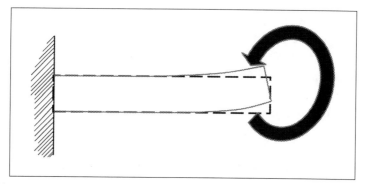

Figure 4.4: Beam with one fixed end subjected to pure bending

A somewhat similar condition exists for a length of shaft subjected to a pure torsion load. The shear stresses distributed on the cross-section will produce shear strains which are angular displacements. The net effect is to produce a uniform twist throughout the length of the shaft. If a straight line were to be drawn on the shaft as in Figure 4.5, this line would be displaced upon load application and would finally occupy the position indicated by the dotted line. The helix angle of displacement would depend upon the shear strains developed at the surface of the shaft.

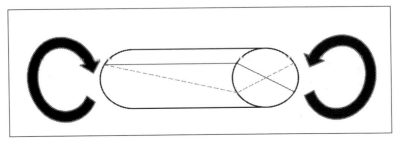

Figure 4.5 Shaft subjected to pure torsion

Only two factors determine the strain for a particular material subjected to stress. One is the magnitude of stress and the second is the type of material and its characteristic stiffness. The actual amount of deflection of a loaded structure will depend on the physical arrangement of the structure and the cumulative effect of the local strains existing in various components of the structure.

Material Constants:

Stress-strain Laws, such as Hooke's Law, use material constants to relate stresses and strains. The constants in Hooke's law are: 1) E called, "Young's Modulus", or the "Modulus of Elasticity" indicating how much longitudinal strain (stretch) per stress? 2) n Called "Poisson's Ratio", the negative of the ratio of the transverse strain (a contraction) to the longitudinal strain (an extension) in an axial tension test. This indicates how much transverse strain per longitudinal? 3) a Called "Coefficient of thermal expansion". This indicates how much longitudinal strain per degree temp change? 4) G = "Shear Modulus". G = E/2(1+υ). Some representative values of material constants for aero structural materials are given below:

Material	Density lb/in3	E lb/in2	G lb/in2	ν	α in/in. F/
Aluminum	0.100	10×10^6	4×10^6	0.25	13×10^{-6}
Steel	0.282	30×10^6	11×10^6	0.36	6×10^{-6}
Titanium	0.162	17×10^6	6.5×10^6	0.31	5.5×10^{-6}
Magnesium	0.065	6.5×10^6	2.4×10^6	0.65	14×10^{-6}

NOTE: Constants for specific alloys and heat-treatments are found in M1L-HBK-5G, Metallic Materials and Elements for Aerospace VEHICLE Structures.

5

STRESS-STRAIN DIAGRAMS
AND MATERIAL PROPERTIES

In order to evaluate the properties of a material and the possible structural application, it is necessary to determine strains corresponding to various levels of applied stress. Laboratory tests are then conducted which subject a specimen of material to various magnitudes of stress while strains are recorded at each stress point. If the stresses and corresponding strains are then plotted on a graph, many useful and important properties of the material may be observed.

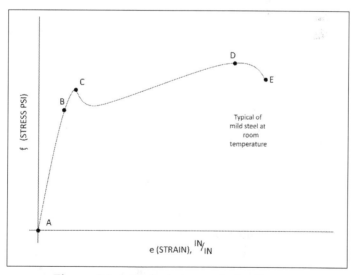

Figure 5.1 A typical stress-strain diagram
for mild steel at room temperature

Figure 5.1 shows a typical stress-strain diagram for mild steel at room temperature. As stress is first applied and then increased, the strain begins and increases in direct proportion to the stress (i.e., the stress-strain diagram is a straight line from A to B). Therefore, if this material were subjected to any stress between A and B, the material would snap back to zero strain upon the release of stress. Because of the elastic nature of the material in this range of stress, the stress range from A to B is referred to as the elastic or proportional range of the material. This definition then implies that stresses up to point B will not cause permanent deformation of the material.

If the applied stress is gradually increased above the value at point B, the plot of stress vs. strain will deviate slightly from a straight line. There will then be some small but measurable permanent strain incurred. Hence, point B is the end of the elastic or proportional range of the material and the value of stress at point B is termed the elastic or proportional limit (σ_p).

Should the stress be gradually increased up to point C, a noticeable yielding of the material will be apparent. At point C, the strain suddenly increases without further increase in stress. In fact, with most ductile steels, there may be a decrease in stress resisted by the material as large plastic strain takes place. It is obvious that any stress above the value at point C will produce very large and objectionable permanent strain. To verify this condition, assume that a stress is applied up to point X on the diagram in Figure 5.2. The strain at this point is quite large. If the stress were to be released, the material would relax along the dashed line (which is parallel to the original straight line from B to A). At point Y, the stress is again zero but a large permanent deformation has taken place. The value of stress at point C is logically termed the yield point or yield stress (σ_y) of the material and is the stress beyond which large and objectionable permanent strain takes place.

The stress-strain diagram of Figure 5.2 does show that the material is capable of withstanding stresses greater than the yield stress, but not without large permanent strains. If the stresses above point C are gradually applied, the material will continue to withstand higher and higher stresses until the very ultimate strength capability is reached at point D. If any attempt is made to subject the specimen to a stress greater than the value at point D, failure will begin and will be complete at point E. Since the value of stress at point D is the highest stress the material can withstand without failure, it is termed the ultimate strength or ultimate stress (σ_u) of the material.

The two most important strength properties which are derived from the stress-strain diagram are the yield strength and the ultimate strength. There is a direct analogy between these two properties and the operating strength limitations of an airframe structure. If the material shown on the stress-strain diagram is never subjected to a stress above the yield point, no significant or objectionable permanent strains will take place. If an airframe structure is never subjected to a load condition greater than the limit load, no significant or objectionable

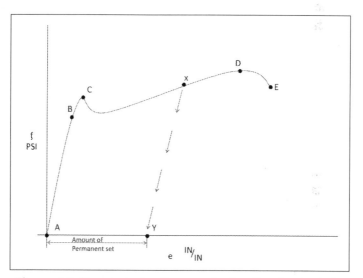

Figure 5.2 Stress-strain diagram for loading beyond the yield point and relaxation from points X to Y.

permanent deformations will be incurred. If the material shown in the stress-strain diagram were subjected to a stress above the yield point, large and undesirable permanent strains will take place. If an airframe is subjected to a load condition greater than the limit load, undesirable permanent deformation of the structure may be anticipated (i.e., permanently distorted fuselage, bent wings, deformed tanks, etc.). If an attempt is made to subject a material to a stress greater than the ultimate strength, failure will then occur. If any flight condition is attempted which produces loads greater than ultimate load, actual failure of the airframe is imminent. The basic stress-strain diagram of Figure 5.3 readily defines five of the most important static strength properties of a material.

1. The elastic or proportional limit is the end of the elastic region of the material. A part subjected to stresses at or below the proportional limit will experience no permanent deformation. Upon release of stress, the part will snap back to the original unstressed shape.

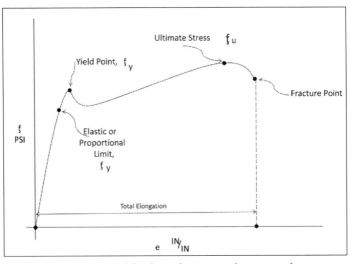

Figure 5.3 Critical static strength properties
of a typical stress-strain diagram

2. The yield strength is the highest practical value of stress to which a material should be subjected. Stresses between the proportional limit and the yield point will cause only slight and barely measurable permanent deformation. Any stress above the yield point will result in large and objectionable permanent deformation.

3. The ultimate strength is the maximum stress which a material can withstand without failure. Extremely large and undesirable permanent deformation will ordinarily result when this point is approached.

4. The fracture point or the effective stress at time of failure is determined primarily to evaluate the manner of fracture and the ductile quality of the material.

5. The total strain or total elongation of the material at the point of fracture is an indication of the ductility of the material. Any metal which would have less than five percent elongation in a two inch test length is considered to be brittle, ordinarily too brittle to be applicable to an ordinary aircraft or missile structure.

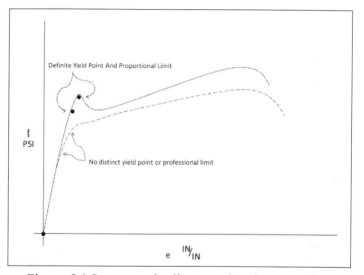

Figure 5.4 Stress-strain diagram showing materials
with and without a distinct yield point

At this point, it is appropriate to present some of the variations in stress-strain diagrams due to manner of loading or material type. One point to consider is that stress-strain diagrams are not usually used in connection with shear properties of a material. The proportional limit, ultimate strength, etc. in shear are not true properties because of section or form considerations. Cross-section dimensions and area distributions will cause significant variations in the shear strength capabilities.

Many materials used in aircraft construction (aluminum alloys, magnesium alloys, and some steels) do not exhibit a definite yield point or a distinct proportional limit. Figure 5.4 illustrates this fact. In such an instance, it is necessary to define the yield point and proportional limit as a given departure from the original straight line. The limit of proportionality is then arbitrarily established as the stress which produces an offset or departure from the original straight line of 0.0001 in/in. The yield stress is determined as the stress which produces a departure from the original straight line of 0.002 in/in. In this case, an applied stress equal to the yield stress would result in a permanent set of 0.2 percent. This amount is considered acceptable for ordinary purposes. Figure 5.5 illustrates the offset principles previously discussed.

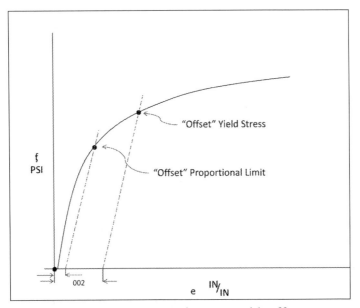

Figure 5.5 Stress-strain curve with offset
yield stress and proportional limit

Compression stress-strain diagrams are similar to tension stress-strain diagrams except that the departure from proportionality generally occurs sooner and more gradually. Compression stress-strain diagrams are more difficult to obtain correctly because of buckling of the specimen. As compression buckling of a specimen constitutes a failure, it is obvious that compression ultimate strength is not a true property and can not be determined as a specific quality of a metal. The stiffness of a metal and the physical arrangement of the structure combine to determine the stability of the structure when subjected to compression loads. Figure 5.6 compares stress-strain curves under tension and compression loads.

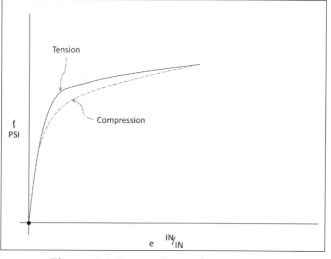

Figure 5.6 Comparison of tension and
compression stress-strain curves

In order to determine other material properties expressed by the stress-strain diagrams, a closer examination must be made of certain areas of the diagram. By an inspection of the straight line portion of the stress-strain diagram, an evaluation may be made of the inherent stiffness of the material.

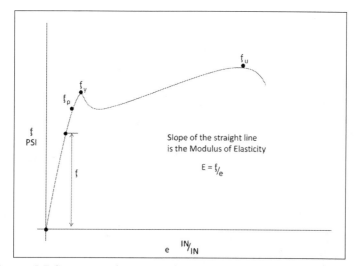

Figure 5.7 Stress-strain curve showing the modulus of elasticity

In the elastic range of a material, the proportion between stress and strain is the Modulus of Elasticity (sometimes called Young's Modulus) and the magnitude of this proportion is a direct measure of the inherent stiffness. A high value for the Modulus of Elasticity will indicate a very stiff or rigid material, while a low value indicates low stiffness or greater flexibility.

When the stress-strain diagrams for three different materials are compared as in Figure 5.8, the difference in the slopes and the proportions of inherent stiffness are readily apparent.

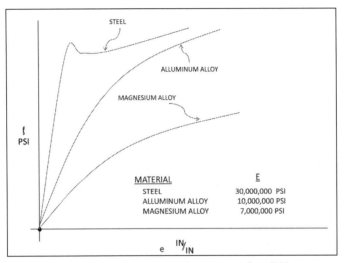

Figure 5.8 Comparison of stress-strain curves for different materials

One important point to consider is that the modulus of elasticity (E) cannot be altered by heat treatment. Thus, the modulus of elasticity is an intrinsic property of the type of metal. Figure 5.9 shows the typical stress-strain diagrams for steel in various conditions of heat treatment. Notice that the origin of each has the same slope. While the strength and ductility are changed by heat treatment, the stiffness of the elastic material remains unaltered. Increasing hardness by heat treatment will simply increase the stress at which plastic flow begins. Thus, if excessive elastic deflections of a part exist, the problem will not be solved by heat treatment or changing alloys.

Only two solutions exist: (1) lower the value of the operating stress or (2) change to a material type which has a higher elastic stiffness (higher E).

Actually, the elastic modulus of a given metal will vary only with temperature, as elevated temperatures produce a lower modulus of elasticity. For example, a high strength aluminum alloy at 600°F will exhibit a modulus of elasticity which is only one-half the value shown at room temperature.

The elastic range of a material is the area of most general interest for ordinary structural investigation. However, when investigating the phenomenon of buckling of short columns and the behavior of structures at loads near ultimate, the stiffness of a metal in the plastic range is of particular interest. As was previously noted, the proportion between stress and strain is defined as the modulus of elasticity.

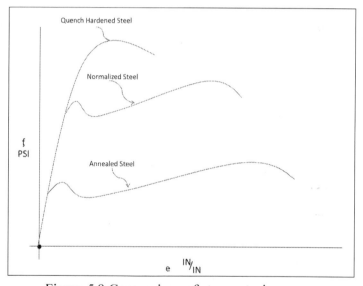

Figure 5.9 Comparison of stress-strain curves
for steel under different heat treatment

At any stress below the proportional limit, this proportion between stress and strain is of a both fixed and constant value. Above the proportional limit, this definition results in a proportion which is known as the *secant modulus*. Once beyond the proportional limit, the proportion between stress and strain will noticeably decrease. Hence, the secant modulus will have a value lower than the modulus of elasticity. Another method of measuring the stiffness in the plastic range is the slope of a line drawn tangent to the stress-strain diagram at some stress above the proportional limit. The slope of this tangent line defines a value known as the *tangent modulus*. The tangent modulus then measures an instantaneous stiffness while the secant modulus measures a gross or cumulative stiffness. Figure 5.10 gives the procedure of calculation of these properties and illustrates the typical variations of the secant modulus and tangent modulus for an aluminum alloy.

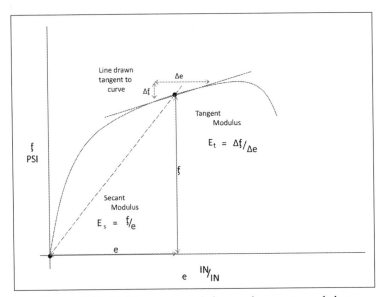

Figure 5.10 (a) Tangent modulus and secant modulus
calculations from stress-strain curve

The loss of stiffness in the plastic range is appreciated when it is realized that small changes in stress will produce larger changes in strain than in the elastic region. The energy storing and energy absorbing characteristics are of particular importance in defining the properties of a material. Any stress produced in a material requires the application of a force through a distance and as a consequence, a certain amount of work is done. Consider a cubic inch element of steel with a gradually applied stress of 30,000 psi (as in Figure 5.11).

As the stress is gradually increased from zero to 30,000 psi, the strain gradually increases from zero to:

$$\varepsilon = \frac{\sigma}{E} = \frac{30,000\,psi}{30,000,000\,psi} = 0.001\frac{in}{in}$$

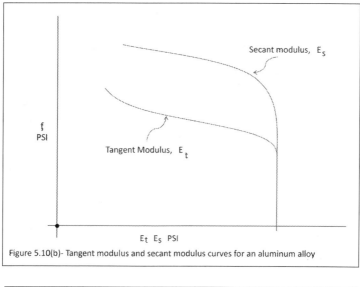

Figure 5.10(b)- Tangent modulus and secant modulus curves for an aluminum alloy

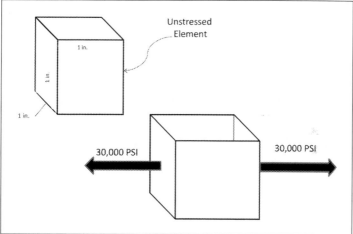

Figure 5.11 Cubic steel element with applied stress

The work done during this gradual stressing is the product of the average force and the distance:

$$Work = (F_{avg})(d) = \frac{30,000\,psi}{2}(0.001\frac{in}{in}) = 15\frac{in-lb}{in^3}$$

Thus, 15 inch-pounds of work would be required to produce the final stress of 30,000 psi in the cubic inch element. As shown in Figure 5.12, this amount of work is actually represented by the area enclosed on the stress-strain diagram at this particular stress level. This area principle is then used with the stress-strain diagram to describe the energy properties of a material. If a material is stressed to the proportional limit, all work done in producing this stress is stored elastically in the material. Therefore, the area under the straight line portion of the stress-strain diagram is a direct measure of the energy storing capability and is referred to as the *modulus of resilience*. Of course, the manner of loading (tension, compression, shear, etc.) must be specified since the work done will be different depending on the manner of loading. If a material were stressed all the way to failure, the entire area under the stress-strain diagram is a measure of the work required to fail the material. The toughness or energy absorbing quality of the material is evaluated in this manner.

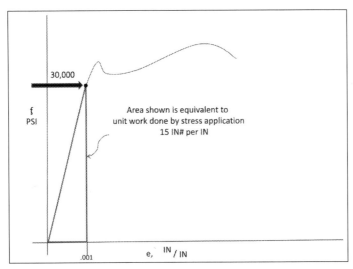
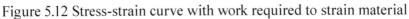

Figure 5.12 Stress-strain curve with work required to strain material

5.1 Ductile and Brittle Materials:

A comparison of the stress-strain diagrams for ductile and brittle steel should define the properties of resilience and toughness. The shaded areas shown in Figure 5.13 denote the energy storing capabilities for ductile and brittle steels. The brittle material has the higher elastic limit and consequently a higher modulus of resilience.

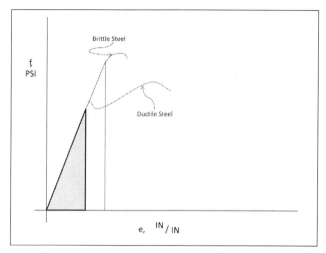

Figure 5.13 Comparison of brittle steel and ductile elastic limits

The shaded areas of Figure 5.14 denote the energy absorbing capabilities of the same materials. The ductile material, while having lower strengths, develops much greater strains. As a result, the ductile material will require a greater amount of work to produce failure and is therefore the tougher material. The additional work required to break the softer, more ductile material goes into forcing the metal particles to slip relative to each other.

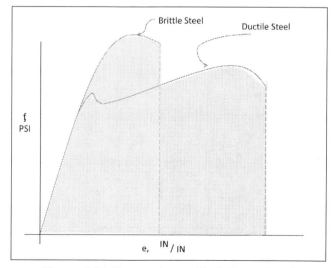

Figure 5.14 Comparison of brittle steel and
ductile steel energy storing capabilities

When a specimen of ductile material is fractured during a laboratory test, this effect may be appreciated by handling the broken specimen immediately after failure. A ductile specimen will be warm to the touch as the work absorbed by the material is converted into heat.

It may seem strange that it is possible to fail the more ductile metal at a lower stress, but still require that more work be done. If so, remember that work is the product of average force and distance.

It would be desirable for a structural material to have a high yield point, high resilience, high ultimate strength, and also high toughness. However, materials which have very high strength usually have low ductility and low toughness. The balance between the strength requirements and toughness (or energy absorbing) requirements will depend on the particular application in a structure. As examples of the two extremes, consider (1) a reciprocating engine valve spring and (2) a protective crash helmet or protective headgear. In the fabrication of a valve spring, a material must be selected which is of very high strength and has great resilience. Such a material would necessarily have low toughness, but in such an application, toughness is unimportant and resilience is given primary consideration. In the construction of a true protective crash helmet or hard hat, sharp impact blows first

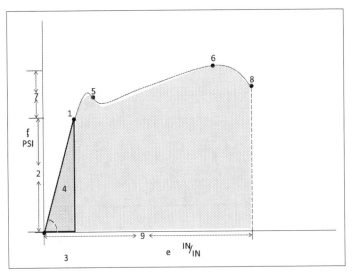

Figure 5.15 Stress-strain diagram showing
ten important characteristics.

must be distributed, then the energy of impact absorbed as far as possible. This requires a thickness of a crushable material for the inner lining which has high toughness per unit weight. Actual strength is not necessarily a factor since energy absorption requirements predominate.

The stress-strain diagram of a material will then furnish all necessary information to determine the static strength properties. Actually, there are ten important points of information that may be gained from an inspection of the stress-strain diagram. As shown in Figure 5.15, these points are as follows:

(1) The proportional limit is the stress which denotes the end of proportionality between stress and strain. If not clearly defined, an offset of 0.0001 in/in is used.

(2) The elastic range of stress is the range of stress and strain up to the proportional limit.

(3) The modulus of elasticity is the slope of the straight line portion of the stress-strain diagram. This slope measures inherent stiffness. Beyond the proportional limit, the secant or tangent modulus will be appropriate.

(4) The resilience is measured by the area under the straight line portion of the stress-strain diagram. This indicates the ability to store energy elastically.

(5) The yield stress is the value of stress above which objectionable amounts of permanent strain are incurred. If not clearly defined, an offset of 0.002 in/in is used.

(6) The ultimate strength is the largest stress that the material can withstand without failure. This represents the maximum load-carrying capability for static loads.

(7) The plastic range of stress is between the proportional limit and the ultimate stress. Permanent strains occur in this area. Below the yield stress, these permanent strains are relatively small and insignificant. Above the yield stress, these permanent strains are large and objectionable.

(8) The fracture point is the effective stress at time of failure. It is noted primarily to evaluate the manner of failure and the ductile quality of the material.

(9) The total elongation is a measure of ductility. If elongation is less than five percent in a two inch specimen length, the material is considered brittle.

(10) The toughness of a material is represented by the total area under the stress-strain diagram. This indicates the amount of work required to fail the material and denotes energy absorption capability.

5.2 Work Hardening or Strain Hardening:

An interesting phenomenon in connection with the stress-strain diagram is work hardening or strain hardening. Suppose that a material is subjected to a stress beyond the yield point and then released. If stress is subsequently reapplied, the new stress-strain diagram will be a straight line up to the point where stress was released and then will continue along the original curve. Figure 5.16 illustrates this process. The material, which has been permanently stretched, will have higher proportional and yield strengths.

Many of the typical aircraft materials (aluminum alloys, stainless steels, etc.) are work hardened to produce these beneficial gains in strength properties. Of course, the work hardening must be limited to prevent loss of ductility and formation of flaws and fissures.

The effects of ductility and the plastic range of stress are quite significant in predicting the failing load of a structure. The ultimate stress has been defined previously as the maximum stress a material will withstand without failing. This stress is designated as the tension ultimate (σ_u) or shear ultimate (τ_u) depending on the manner of loading and is determined by tests of small specimens. If these values of strength are used to predict the ultimate strength capability of large sections in bending and torsion, noticeable errors may result.

An elastic stress distribution in bending may be predicted by the following relationship:

$\sigma_b = \dfrac{My}{I}$, where, σ_b is bending stress [psi]; M is applied bending moment [in-lb.]; y is element distance from neutral axis [in]; and I is moment of inertia of the cross-section [in⁴].

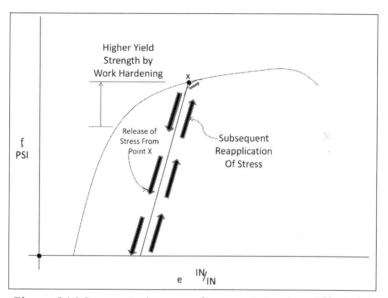

Figure 5.16 Stress-strain curve for a work-hardened material

Such a stress distribution is linear, varying directly with the element distance from the neutral axis. A typical elastic stress distribution is shown in Figure 5.17.

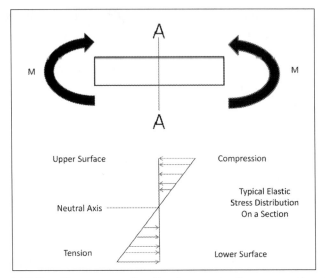

Figure 5.17 Bending stress distribution up to the proportional limit

Whenever stresses are produced which are beyond the proportional limit of the material, the bending stress distribution is no longer linear. Actually, the bending strain distribution tends to remain linear, but due to the loss of proportionality between stress and strain in the plastic range, the stress distribution is nonlinear. Figure 5.18 shows a typical stress distribution resulting from bending loads which create stresses in excess of the proportional limit.

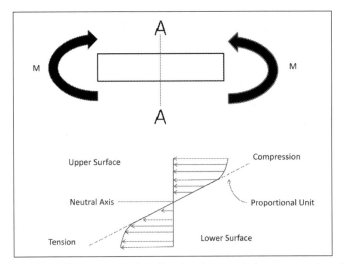

Figure 5.18 Bending stress distribution beyond the proportional limit

Notice that for a given maximum stress at the outer fiber, the inelastic bending stress distribution requires a greater applied bending moment than the elastic stress distribution. The reason for this is that the outer fibers begin to yield, allowing the underlying fibers to develop a stress higher than predicted by elastic theory. Thus, the use of the equation

$$\sigma = \frac{Mc}{I}$$

to compute the maximum bending stress is valid only for maximum stresses which do not exceed the proportional limit. In order to predict the failing load of a structure in bending, the same form of equation may be used with a fictitious ultimate stress referred to as the bending modulus of rupture (R_b). This bending modulus of rupture is defined by the following equation:

$$R_b = \frac{M_b c}{I}$$

Where R_b = bending modulus of rupture [psi]
               M_b = bending moment to cause failure [in-lb]
               c = distance to critical outermost element [in]
               I = section moment of inertia [in^4]

If the material is ductile and no buckling of the section occurs, the bending modulus of rupture will be some value greater than the tensile ultimate strength. In a perfectly ductile, stable cross-section, the bending modulus of rupture could be 1.5 times the tensile ultimate strength. In a very brittle stable cross-section, the bending modulus of rupture would be equal to the tensile ultimate strength. If the cross-section is composed of thin-walled unstable elements which are apt to buckle, the bending modulus of rupture may be much less than the tensile ultimate strength. The bending modulus of rupture is obviously dependent upon the type of material (especially ductility) and the shape or form of the cross-section.

An analogous situation exists for sections subjected to torsion loading and is due to the form of the cross-section and the character of the material. In order to predict the failing torsion load of a structure, another fictitious stress referred to as the *torsion modulus of rupture* (R_T) is used. This torsion modulus of rupture is defined by the following equation:

$$R_T = \frac{T_u c}{J}$$

Where R_T = torsion modulus of rupture [psi]
  T_u = torque to cause failure [in-lb]
  c = distance to critical outermost element [in]
  J = polar moment of inertia of cross-section [in^4]

If a metal is subjected to a stress in the plastic range, the strain will continue to change in the direction of stress although the stress is held constant. This is a condition which is most common to metals at high stresses and high temperatures and is known as *creep*. Figure 5.19 is a typical creep curve for a metal at a constant stress and temperature.

Upon application of stress, a high initial creep rate will develop, and then decrease to the minimum value of steady creep. During the second stage, the creep rate is essentially constant. After a period of time, the third and final stage of creep will take place with an increase in creep rate and final rupture. Since the stress to cause rupture varies inversely with time and temperature, creep problems are of greatest importance in light weight, high temperature structures.

In the design of high temperature structures, the amount of deformation allowable may be a more severe criterion than the actual rupture strength. This would occur when excessively strained components would not function properly or would fail at loads lower than normally anticipated.

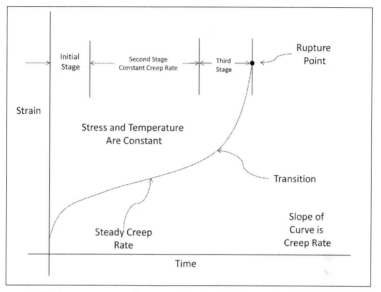

Figure 5.19 Creep curve for metal at constant stress and temperature

Stress-strain diagrams for a material may be greatly altered when the rate of stressing is very high. When stress is applied very suddenly (almost instantaneously), the result is impact stresses.

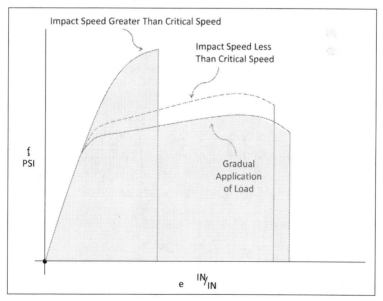

Figure 5.20 Comparison of stress-strain curves
under different loading conditions

The amount of energy absorbed under impact conditions may be significantly different from the energy absorbed when the load is applied steadily and gradually. The actual speed or rate at which a material is stressed will determine what changes in energy absorption take place. In the case of an ordinary structural material, an increase in the rate of stressing (above that of very gradual load application) will initially produce a slight increase in the energy absorbed. With continued increase in rate of stressing, a critical speed will be reached and the energy absorbed will be a maximum. Above this speed, the toughness will be greatly reduced. If the stress-strain diagrams for impact stressing were recorded and compared with the stress-strain diagram obtained from gradual load application, the result would be similar to Figure 5.20. When stress is applied above the critical speed, the failing stress would be higher but the elongation would be greatly reduced. In fact, the effect of impact stresses beyond the critical speed is to produce brittle type failures in tough or ductile materials. On the other hand, a brittle material will not show any great effect of high stress rates since there is very little ductility or energy absorbing capability for gradually applied loads.

While impact stresses near the critical speed are not ordinarily encountered in airframe structures, due consideration must be given to the case of dynamic machinery and mechanisms. The effect of impact stresses is most important when there are severe discontinuities in the shape of a part or when a part is operated at low temperatures.

5.3 Commonly Used Terms:

This section gives engineering definitions of longitudinal and/or shear strains. It gives corresponding diagrams and enables one to draw from memory stress-strain diagrams for materials with and without well-defined yield points. Given the appropriate stress-strain diagram, it helps to label the following points on the diagram:

Proportional limit (elastic limit), Yield point

Ultimate stress

Fracture stress

Offset proportional limit (know the amount of the offset)

Offset yield point (know the amount of the offset)

Describe strain hardening.

Describe the effects of impact loading on the stress-strain curve

Define the following terms:

i1. Elastic

i2. Engineering stress

i3. Plastic

Use a stress-strain diagram to explain the following terms:

j1. Modulus of Elasticity (Young's Modulus)

j2. Plastic and elastic strains

j3. Toughness vs. strength

j4. Ductile vs. brittle

Discuss creep:

k1. The conditions, which cause it

k2. The four (or three) stages of the creep process

k3. The difference between stress rupture and simple fracture

k4. Discuss the implications of the creep process on the proper use of engines during normal operation and emergency conditions, including observance of operating limitations.

5.3.1 Material Constants

STRESS-STRAIN LAWS, such as Hooke's Law, use material constants to relate stresses and strains. The constants in Hooke's law are

E = "Young's Modulus", or the "Modulus of Elasticity" indicating how much longitudinal strain (stretch) per stress?

n = "Poisson's Ratio", the negative of the ratio of the transverse strain (a contraction) to the longitudinal strain (an extension) in an axial tension test. This indicates how much transverse strain per longitudinal strain?

a = "Coefficient of thermal expansion". This indicates how much longitudinal strain per degree temp change?

G = "Shear Modulus" = $G=E/2(1+v)$

Some representative values of material constants for aero structural materials are given below

Material	Density lb/in3	E lb/in2	G lb/in2	v	α in/in. F/
Aluminum	0.100	10×10^6	4×10^6	0.25	13×10^{-6}
Steel	0.282	30×10^6	11×10^6	0.36	6×10^{-6}
Titanium	0.162	17×10^6	6.5×10^6	0.31	5.5×10^{-6}
Magnesium	0.065	6.5×10^6	2.4×10^6	0.65	14×10^{-6}

NOTE: Constants for specific alloys and heat-treatments are found in MIL-HBK-5G, Metallic Materials and Elements for Aerospace VEHICLE Structures Melting points of metals are NOT found in MIL-HBK-S.

5.3.2 Stress strain Diagram:

For ductile material with identifiable yield point for elastic vs. plastic strains/regions a stress-strain diagram is given below:

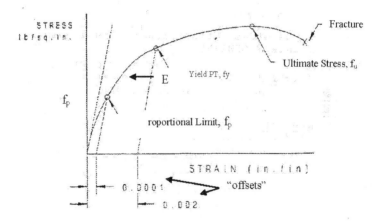

And Stress- strain diagram for ductile material with no clear-cut yield point is given below:

5.3.3 Ductile versus Brittle Materials:

"Ductile" implies that total strain to failure is equal to or greater than 5% (where 5 is an arbitrary number).

"Brittle" implies that total strain to failure is equal to or less than 5%.

But both ductile and brittle tend to be used as comparatives.

5.3.4 Strength versus Stiffness

"Strength" is the ability to withstand stress referenced to ultimate strength.

"Toughness" is the ability to absorb energy prior to failure.

The area under stress-strain diagram is a quantity called Strain Energy Density.

$$\frac{lb}{in^2} \times \frac{in}{in} = \frac{inlb}{in^3}$$

Where "in lb" is energy and "in³" is volume.

Modulus of Toughness is total area under the curve, out to fracture.

A comparison of two materials is given by the following figure:

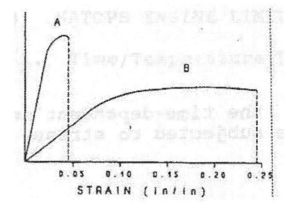

"A" is: stronger, brittle and stiffer

"B" is: tougher and ductile

….and that's the way the world works---these properties naturally go together.

3.3.4 Effect of Impact Loading:

The impact load changes the behavior of ductile to brittle. In most materials an increase in strain rate will cause an increase in the effective modulus of elasticity and an increase in the fracture stress. The following figure demonstrates this transition. In most materials an increase in strain rate will cause an increase in the effective modulus of elasticity and an increase in the fracture stress

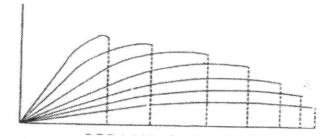

"Critical strain rate" is the strain rate at which the energy absorbed before failure is a maximum.

Important Fact: Ductile materials respond in brittle fashion when they are strained faster than their critical strain rate.

Impact is probably the definitive case of rapid load onset, i.e., rapid strain rate. Thus impact loads cause brittle Failure. Low temperatures will also cause a ductile to brittle transition (DTBT).

5.3.5 Creep Response of Materials:

Definition: CREEP is the time-dependent deformation produced in solids subjected to stress. A representative creep diagram of materials is given below:

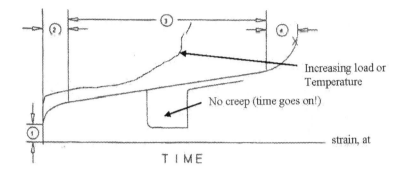

Increasing load or Temperature

No creep (time goes on!)

strain, at

T I M E

(1) Initial Stage: Initial elastic, plastic (if any), and thermal strain, at time zero.

(2) Second Stage: Transient stage in which creep **rate** decreases with time due to strain hardening.

(3) Third Stage: <u>Constant Rate Creep.</u> Strain hardening is balanced by <u>annealing</u> **(softening) due** to elevated temperature.

(4) Fourth Stage; Increasing **creep rate,** necking, **leading to..**

(5) Stress Rupture: **Fracture after a period of time at constant load and temperature.**

6

EFFECT OF HEAT TREATMENT
AND COMPOSITION

Most materials used in aircraft and missile structures must be heat treated to obtain the desired physical characteristics. In order to understand the means by which heat treatment and composition affect the physical properties of a material, the general composition of a metal must be considered. The metallic atom is an extremely small particle and is not visible by ordinary optical means. This metallic atom is composed of a positively charged nucleus and negatively charged electrons which revolve about the nucleus. The electrical forces of attraction between the nucleus and the revolving electrons are balanced by the centrifugal forces due to rotation. The electrons of each metallic atom are attracted by the nuclei of adjacent atoms but repelled by the electrons of these adjacent atoms. When the atoms are set in a given arrangement, as in a solid metal, the interatomic forces determine the theoretical strength and stiffness of the metal.

When a metal is in the molten state, the metallic atoms have no fixed arrangement or order. When the metal is cooled to the solid state, the atoms begin to group themselves in a definite pattern or framework known as a space lattice. Hence, any metal in the solid state will appear as a crystalline substance, with the crystalline form being the result of metallic atoms arranged to a given space lattice. Ordinary metals will form one of the following types of space lattice: (1) body centered cubic (2) face centered cubic or (3) close packed hexagonal. Figure 6.1 illustrates each of these three lattice forms.

Upon cooling, most ductile metals form a face centered cubic lattice while most brittle materials form a body centered cubic lattice. Aluminum, copper, gold, nickel, silver, and gamma iron form face centered cubics while chromium, molybdenum, tungsten, and alpha iron form body centered cubics. To appreciate the structural arrangements of a solid metal, inspect Figure 6.2. If the surface of a metal sample were polished and magnified, a granular appearance would be noted. The grains are made up of the crystals or crystalline fragments which are the product of the atomic latticework. The intrinsic strength properties of a metal are determined by the arrangement and spacing of this atomic lattice structure. However, the actual strength properties of this metal will be greatly influenced by the distribution of dislocations in the lattice, discontinuities in the crystals or crystal fragments, and the grain boundary properties and defects.

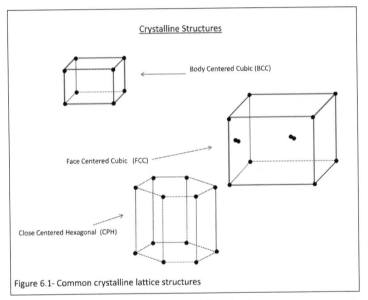

Figure 6.1- Common crystalline lattice structures

Figure 6.1 Common crystalline lattice structures

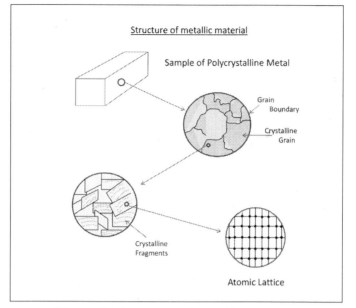

Figure 6.2 Structure of metallic material

When a molten metal begins to solidify, crystallization will begin simultaneously at many random points with each point being a center for progressive crystallization. As the metal continues to solidify, the crystalline pattern extends until boundaries are formed between adjacent areas of crystalline growth. At this point, crystallization is complete and the metal is composed of an extremely large number of crystalline grains, each denoting a separate origin of crystalline growth.

The typical pattern of crystallization and crystalline grain development is illustrated in Figure 6.3. It should be noted that the origins of crystalline development occur in a very random pattern with no immediate orientation to adjacent sources of crystallization. Therefore, as the crystalline grain development is completed, small boundary areas will not be able to arrange themselves into the orderly lattice structure and must settle into some unusual or strained arrangement. At ordinary temperatures, the atoms in the grain boundaries will be subject to great strain as will the crystalline material immediately adjacent to the grain boundaries. Because of the existing great internal strain, the strength and hardness of the grain boundary material will appear to be greater than that of the crystalline grain material. For this reason, fine-grained structures generally have higher strength and hardness than coarse-grained structures. The inference is that the fine-grained structure contains more irregular grain boundary areas.

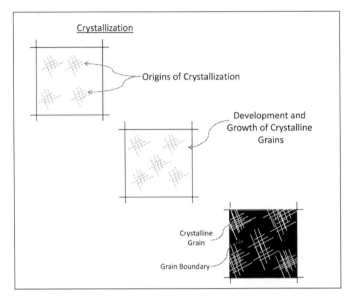

Figure 6.3 A typical pattern of crystallization and crystalline grain

The relationship of strained atomic arrangement with strength and hardness is an extremely important factor in the heat treatment of metals. Any process or environment which varies the strain of the atomic arrangement in the grain boundary or within the grain or crystal will affect the stiffness, strength, and hardness. Thus, character of the metal, rate of cooling, alloying, ingredients, impurities, operating temperature, cold or hot working, etc. may produce significant changes in the physical properties of a metal. An example of these effects may be seen in the case of the grain boundary areas. At normal temperatures, the boundary atoms are subject to great strain and the grain boundaries are quite strong. At elevated temperatures, these atoms have greater mobility and strain is relieved. The grain boundary areas are then the weakest part of the structure and can be easily separated at excess temperatures. This condition could partially account for the creep of metals at high temperatures. If alloys or impurities exist in a metal, these foreign atoms may be pushed ahead into the grain boundary upon crystallization. If these atoms increase the strain of the atomic arrangement in the grain boundary, strengthening will result. If these atoms decrease the strain of the atomic arrangement, weakening will result.

The properties of steel which can be derived by heat treatment depend to a great extent upon the composition of the steel. Carbon is the principal element in steel other than iron. Hence, any steel which contains only iron and carbon not in excess of 1.7% (plus other unavoidable impurities) is referred to as *plain carbon steel.*

If other elements such as nickel, molybdenum, chromium, vanadium, silicon, etc. are purposely included in addition to the carbon, the steel is termed an *alloy steel.* Steel with a carbon content of 15% to 30% is referred to as *low carbon steel* and is useful in applications where ductility is necessary and strength is relatively unimportant. If the carbon content is from 30% to 60%, the steel is a *medium carbon steel* and is useful for various machine parts requiring considerable strength and toughness. Steel containing more than 60% carbon is termed *high carbon steel* and is used for such parts as springs and cutting tools where great strength and hardness are required and reduced toughness and ductility can be tolerated. *Cast iron* is an alloy of iron with carbon in excess of 2.0%. Such a metal is very hard and extremely brittle at any ordinary temperatures.

The heat treatment of steel involves particular control of time and temperature. If plain carbon steel is heated to 1350-1400°F, a change will take place in the internal structure. At this critical temperature, the carbon will go into a solid solution with the iron producing a structure known as *austenite.* This transformation to the solid solution is the first step necessary in the heat treatment of steel and the subsequent cooling of the transformed metal will then determine the physical properties. The most important factor is the rate at which the metal is cooled. If cooling is accomplished very slowly as in a furnace or insulating material, a very coarse-grained structure is produced which is primarily *pearlite* (iron and iron carbide). This coarse-grained structure will be quite soft and very low strength. Such is the state of steel after annealing. If the rate of cooling from the critical temperature is increased by allowing the part to cool in still air, the part is said to be normalized. This greater rate of cooling will produce a more finely dispersed pearlite structure.

Som R. Soni

Such a fine-grained structure would be harder and have higher strength than the annealed material.

Finally, if the steel is cooled very rapidly from the critical range by a quench in cold water, a needle-like structure called *martensite* is produced. Martensite is an extremely hard, high-strength but brittle structure and generally is unmachinable except by grinding. This form of steel may contain high internal stresses from the rapid quench and is so brittle that it is usually necessary to temper steel after quenching.

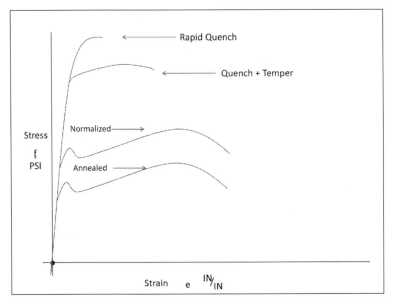

Figure 6.4 Stress-strain diagrams for various heat treatment processes

Tempering of steel is accomplished by reheating to some temperature below the critical range and maintaining this temperature for a period of time. This process results in a gradual transformation of the needle-like martensite to a more granular form and restores toughness and ductility. The results of the various processes are indicated by the stress-strain diagrams of Figure 6.4.

One item basic to the control of the heat treatment process is a *Time-Temperature-Transformation* diagram which will denote the beginning and end of the austenite transformation. One of the diagrams for a plain carbon steel of 0.80% carbon content is shown in Figure 6.5. Due to the general shape of the transformation lines, the diagram is usually referred to as the *S curve*. The left hand edge of the shaded band denotes the beginning of transformation while the right hand edge of the shaded band denotes the end of transformation.

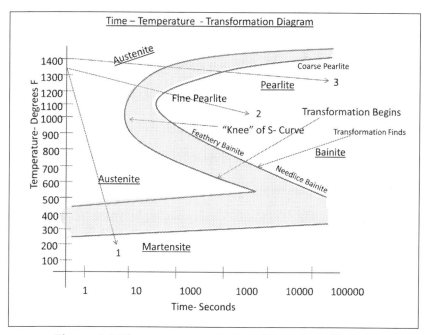

Figure 6.5 Time-temperature-transformation diagram

If various rates of cooling are described by lines 1, 2, and 3, the S curve will define the products of the transformation process. Line 1 describes the very high rate of cooling required to cool past the knee of the S curve. The transformation is complete at 250°F and the result is the extremely hard brittle martensite. Line 2 describes a more gradual cooling process which intersects the S curve after the knee. The transformation is complete at 1100°F and the product is tough, ductile, fine pearlite. Line 3 describes a very low rate of cooling and the product of transformation is the very soft, ductile, coarse pearlite.

If steel is being heat treated to obtain high strength, it is necessary to provide an initial rapid cooling to obtain a transformation which has inherent strength. If a sudden quench (such as shown by line 1 in Figure 6.5) is used, the martensite produced may be so hard and brittle and contain such high internal stresses as to have limited application in practical structures. In order to modify the steel to a more practical form, tempering is usually required subsequent to the quench. Figure 6.6 illustrates several of the methods of tempering.

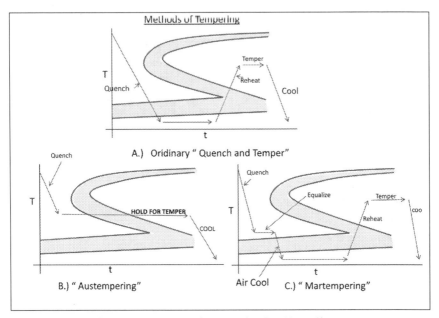

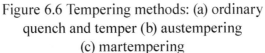

Figure 6.6 Tempering methods: (a) ordinary
quench and temper (b) austempering
(c) martempering

Many elements may be added to iron and carbon to produce *alloy steels*. The addition of these alloying elements will produce many significant changes in the physical properties of steel. The addition of only 3% nickel increases strength and toughness, raises the proportional limit, increases corrosion resistance, and improves the heat treatability of steel. Heat treating of nickel alloy steels can be done at lower temperatures and lower cooling rates are possible in the hardening process. Thus, the possibility of warping and cracking during the quenching process is greatly reduced. If the nickel content is quite high (about 20%), very significant changes will take place: the steel will not transform from austenite, the corrosion resistance is so high that the steel becomes stainless, and the steel becomes non-magnetic. Steels with high nickel content also have very favorable properties at high temperatures (e.g., resistance to oxidation). Nickel steels have great application to machine parts where shock and fatigue loads are important. Actually, the most extensive use of nickel is in conjunction with other alloys such as chromium and molybdenum.

Chromium added to steel increases hardness, strength, proportional limit, and allows much greater penetration of hardness from the surface to the core than is possible with ordinary steels. When chromium is added to steel, the compound of chromium carbide is formed and this adds tremendous wear resistance to steel. For this reason, chromium steels have great application to ball and roller bearing manufacture. If the chromium content of the steel is quite high, the effect is similar to high nickel content as the corrosion resistance is greatly increased and there is great resistance to oxidation at high temperatures. Both nickel and chromium are added to steel for typical machine parts such as propeller shafts, gears, connecting rods, etc. Nickel and chromium combined provide increased tensile strength, fatigue strength, toughness, and ductility. If nickel and chromium are added in certain proportions (18% chromium and 8% nickel), a stainless steel is produced.

Other alloying elements are manganese, tungsten, molybdenum, vanadium, and silicon. These elements may be used in certain proportions to provide various useful properties. Manganese will tend to clean steel by deoxidizing and desulphurizing, and will neutralize or minimize harmful impurities. In addition, manganese provides penetration hardness (i.e., allows more uniform penetration of hardening operations). Tungsten steels have no direct application to aircraft structures, but tungsten is an important ingredient of tool steel and die facings because of the extreme hardness maintained at high temperatures. Molybdenum can be used as a substitute for tungsten in tool steel. Molybdenum is usually used in conjunction with other alloys (chromium and nickel) to improve tensile strength properties at elevated temperatures without sacrifice of toughness at ordinary temperatures. Vanadium is an expensive alloying element which serves two purposes: to deoxidize and to improve shock and fatigue resistance. Vanadium is ordinarily used with chromium to produce the chrome-vanadium steels. Silicon as an alloy will act as a deoxidizer and if used in small quantities, will improve the ductility and impact strength. By the use of these various alloys, the properties of steel may be tailored to most specific structural requirements.

In order to provide a system of designating the steels used in automotive and aircraft construction, the Society of Automotive Engineers (SAE) provided a simple numerical system to identify the composition of ordinary alloy steels. The first digit of this numbering system indicates the type of steel (e.g., "1" indicates a carbon steel, "2" indicates nickel steel, "3" indicates a nickel-chromium steel, etc.). The second digit of this numbering system indicates the approximate percentage of the predominant alloying ingredient. The last two digits indicate the carbon content in hundredths of one percent. By this system of numbering,"2340" steel indicates nickel as the principal alloying ingredient of 3% content and the carbon content is 0.40%. In some special instances, the numbering system is approximate and deviations from the specified system are necessary to describe special steels. Table 6.1 lists common alloy steel numerical designations.

In many instances, it is desirable to produce steel parts which have high strength outer surfaces but maintain a tough ductile interior. This is a special requirement of steel parts with surfaces subject to wear and fatigue type stressing. Such a part is usually machined to shape in a relatively ductile condition and then a hard, outer case is produced by carburizing, nitriding, cyaniding, or induction and flame hardening. Since these processes affect a hard outer case but maintain the tough ductile core, such treatments are referred to as *case hardening*.

Carburizing consists of the introduction of additional carbon into the surface layers of a part. This is accomplished by packing the part in a carbonaceous material and heating to 1600-1800°F. At the elevated temperatures, carbon monoxide acts as the carrier of carbon from the carbonaceous material (e.g., charcoal to steel). A subsequent heat treatment consisting of reheat and quench is required to refine the grain throughout the section and produce the desired properties. The higher carbon content of the surface produces the increase in local hardness and strength. Carburizing can produce a hardened core with as great as 0.125" penetration.

Table 6.1 Common alloy steel designations

Type of Steel and Alloy Content	Numerals and Digits	Type of Steel and Alloy Content	Numerals and Digits
CARBON STEELS		**NICKEL - MOLYBDENUM STEELS**	
Plain carbon	10xx	Ni 1 57 and 1 82 Mo 0 20 and 0 25	46xx
Free cutting	11xx	Ni 3 50 Mo 0 25	48xx
MANGANESE STEELS		**CHROMIUM STEELS**	
Mn 1 75	13xx	Low Cr - Cr 0 27 0 40 and 0 50	50xx
NICKEL STEELS		Low Cr - Cr 0 80 0 87 0 92 0 95 1 00 and 1 05	51xx
Ni 3 50	23xx	Low Cr (bearing) Cr 0 50	501xx
Ni 5 00	25xx	Medium Cr (bearing) Cr 1 02	511xx
NICKEL-CHROMIUM STEELS		High Cr (bearing) Cr 1 45	521xx
Ni 1 25 Cr 0 65	31xx	Corrosion or Heat Resisting	514xx and 515xx
Ni 3 50 Cr 1 57	33xx		
CORROSION OR HEAT RESISTING	303xx	**CHROMIUM - VANADIUM STEEL**	
		Cr 0 80 and 0 95 V 0 10 and 0 15	61xx
MOLYBDENUM STEELS			
Mo 0 25	40xx	**SILICON - MANGANESE STEELS**	
		Mn 0 65 0 82 0 87 and 0 85	
CHROME - MOLYBDENUM STEELS		Si 1 40 and 2 00	
Cr 0 50 and 0 95		Cr 0 17 0 32 and 0 65	92xx
Mo 0 25 0 20 and 0 12	41xx		
		LOW-ALLOY HIGH-TENSILE	950
NICKEL - CHROMIUM - MOLYBDENUM STEELS			
Ni 1 82 Cr 0 50 and 0 80 Mo 0 25	43xx	**BORON INTENSIFIED STEELS**	
Ni 1 05 Cr 0 45 Mo 0 20	47xx	B DENOTES BORON STEEL	xxBxx
Ni 0 55 Cr 0 50 and 0 65 Mo 0 20	86xx		
Ni 0 55 Cr 0 50 Mo 0 25	87xx	**LEADED STEELS**	
Ni 3 25 Cr 1 20 Mo 0 12	93xx	L DENOTES LEADED STEEL	xxLxx
Ni 1 00 Cr 0 80 Mo 0 25	98xx		

Nitriding is another method of case hardening which consists of the introduction of nitrides into the outer surface. This is accomplished by soaking the steel parts in ammonia gas at approximately 1000°F. While nitriding may appear to be more expensive than carburizing, there are certain advantages. Nitriding can be accomplished at lower temperatures and no subsequent heat treatment is required. Thus, there is less chance of warping and distortion, and a higher carbon steel may be used. Also, the nitrided surface is generally of higher hardness and more resistant to corrosion.

Cyaniding is essentially a combination of carburizing and nitriding. Cyaniding is accomplished by soaking the steel parts in molten salts and heated ammonia gas. While cyaniding has the obvious disadvantages of a highly toxic substance, a very thin case can be quickly obtained.

Induction and flame hardening are much simpler processes applied to steels with sufficient carbon content. Each process consists of simple heating and quenching of only the surface to produce the hardened outer case. Induction hardening is often used in the production of reciprocating engine and gear parts. Induction blocks with high frequency, high amperage current set up strong eddy currents which heat the parts in a matter of seconds. A subsequent quench by water spray produces the hardened case.

If a case hardened part is subjected to excessive temperatures in service, there is the possibility that the hardened case will be damaged. A carburized part subjected to extreme temperatures in operation may be decarburized. Excessive wear or fatigue failures may result from the subsequent loss of strength and hardness.

Aluminum alloy is one of the most important metals of general use in aircraft and missile construction. The high strength and low weight of aluminum alloy have made this metal applicable to nearly all parts of airframe construction. Aluminum in the pure state has very low strength and high ductility. While pure aluminum may be strain hardened to some extent, it cannot be hardened by heat treatment. Pure aluminum may have a tensile ultimate strength of approximately 10,000 to 20,000 psi with a minimum elongation of approximately 50%. Although the strength is low, the corrosion resistance of pure aluminum is very high.

Alloying aluminum with certain other metals (particularly copper) will allow heat treatment and the development of tensile ultimate strengths as great as 80,000 to 100,000 psi. Only by alloying is the heat treatment and subsequent hardening possible.

The heat treatment of aluminum alloy depends on the creation of a solid solution of the alloying ingredients and a relationship of decreasing solid solubility with decreasing temperature. Thus, the first step in the heat treatment of an aluminum alloy requires the attainment of a stable, supersaturated solution of the alloying ingredients and the aluminum at elevated temperature. The solution heat treatment is accomplished by heating the alloy to 800-1000°F until the soluble alloying ingredients are brought into a solid solution. The exact temperature and required soaking time depend on the type of alloy and the particular dimensions of the parts.

When sufficient time has been allowed to complete the solution heat treatment, the parts are given an immediate rapid quench and the alloy is held in solution. A solid solution is obtained which is supersaturated and unstable below the solution heat treatment temperature.

The alloy immediately after the solution heat treatment and quench is not yet hardened and at that time has relatively low strength and high ductility. The final step in the hardening operation consists of the decomposition or aging of the unstable solid solution. The extremely fine dispersion of the alloy by the solution heat treatment creates high internal strains in the structure and produces both high strength and hardness of the aluminum alloy.

Some of the aluminum alloys begin a high rate of precipitation hardening immediately after the quench and full hardness is obtained within 48 hours while at room temperature. Certain high strength alloys will have a relatively low rate of hardening at room temperature and may require as long as two months to obtain full hardness. In such cases, an artificial aging may be used to increase the hardening rate by holding the alloy at a temperature of 200-400°F. Thus, the precipitation hardening can be accomplished in a few hours rather than a few months.

Just as heating will accelerate the precipitation hardening the refrigeration can be used to arrest the hardening. If the parts are refrigerated immediately after the quenching operation, the precipitation hardening rate will be very low and the parts will remain relatively soft and ductile. This technique is useful since forming operations may be conducted while the part is ductile and subsequent hardening can take place after the fabrication is complete (e.g., ice box rivets). In fact, forming or straightening operations after the quench have certain advantages. Of course, the possibilities of warping and loss of dimensional tolerances during quench will not be a factor in parts formed subsequent to the quenching operation. Also, the cold working before age hardening reduces the grain size and produces a distinct grain direction. The reduction in grain size is beneficial and the directional properties give added strength along the grain direction.

While pure aluminum has very high resistance to corrosion, the introduction of alloying ingredients creates a problem. Because of the homogeneity within the grain, no great problem exists in this area. However, the conglomeration of elements which separate or precipitate into the grain boundary can create electrolytic cells and incur the possibility of intergranular corrosion. The possibility of intergranular corrosion is greatly increased if the quench during heat treatment is delayed or if the cooling rate is too low. Delayed quench or low cooling rate allows an excess separation of the alloying ingredients and enhances intergranular corrosion.

Protection of the aluminum alloy from intergranular and other forms of corrosion can be effected by various means. One usual method is cladding of the alloy with a thin layer of pure aluminum. Cladding with pure aluminum seals the alloy with the pure metal which has high inherent resistance to corrosion. Since this layer of cladding amounts to approximately 3% to 5% of the plate thickness, considerable care must be taken to prevent damage to the cladding and exposure of the alloy. Welding or heat damage to aluminum alloys may destroy the beneficial effects of prior heat treatment and weaken the metal. In addition, the possibility of inter-granular corrosion is increased.

In summary, it may be stated that the objective of high strength is obtained by some effect which will inhibit plastic deformation. The inevitable consequence is that high material strength will lead to low ductility. The various heat treatments which effect favorable solid solution, precipitation, dispersion, etc. reduce plastic flow within the material. Thus, the practical limitation imposed on any hardening operation is the requirement for ductility, toughness, impact resistance, and shock resistance.

Some particular materials cannot be hardened by heat treatment and must be prepared by work hardening. For example, early forms of stainless steels could be annealed and all favorable properties then obtained by varying degrees of work hardening.

7

BUCKLING

The static ultimate strength of structures subjected to tensile loads is simply a matter of the cohesive strength of the metal. When structures are subjected to compression loads, the ultimate strength is a complex relationship of stability rather than cohesive strength. If a small block of ductile metal is subjected to extreme compression loads, the block distorts and deforms but does not define a true ultimate cohesive strength in compression. In this sense, compression ultimate strength is not definable as a property of the metal. The actual case is that the compression ultimate strength of a typical structure is defined by *buckling*. The mode of failure defined by buckling can be demonstrated by a compression load applied to a long, slender column as depicted in Figure 7.1.

The mode of failure of the long column loaded in compression can be appreciated by the plot of end load versus end deflection of the column with applied compression load. As the compression load is applied and gradually increased, the initial deflection of the column is due to the axial strain of the material. As the application of compression load is continued, a critical load will finally be reached which will cause the column to deflect laterally and no greater load can be maintained. Thus, the compression column can withstand no more than the critical load which produces buckling. This buckling instability defines the maximum, or ultimate, load carrying capability of the compressed structure.

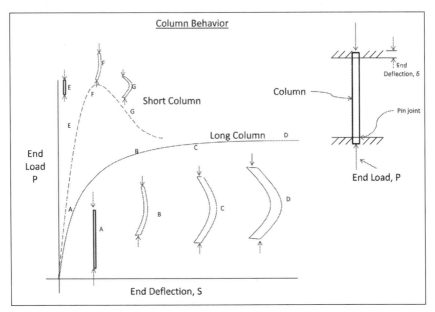

Figure 7.1 Mode of failure of long and short columns

If the column is long and slender and if all stresses remain within the elastic range, release of the critical load will allow the column to snap back to the original unstressed shape. The column would be capable of repeating the same loading procedure and buckling at the same critical load if stresses remain in the elastic range. This type of column is termed the *long column* and it typifies the classic condition of pure elastic stability of a compressed structure.

A *short column* would experience stresses greater than the proportional limit at the critical load and would incur permanent deformation. The short column would not snap back to the original unstressed shape upon release of the critical load and would not be capable of repeating the same loading without failure occurring at a smaller load.

The difference in the mode of failure of the long and short column is illustrated in Figure 7.1. The plot of deflection versus end load for the long column shows that the column is capable of resisting the critical load in the deflected condition. Hence, the failure of the classic long column is a condition of neutral stability. At the critical load, the end deflection may vary but the column maintains the critical load. This condition is true within the limiting conditions of elastic stresses and relatively small deflections.

The case of the short column differs because stresses at or before the critical load are not elastic. Thus, the mode of failure of the short column is typified by a reduction in load with an increase in deflection past the critical loading. In other words, the failure of the short column is unstable and if the point of the critical load is exceeded, the column will collapse. While the short column of a given cross-section is capable of higher critical loads, the mode of failure is a distinct instability.

The strength of a column is actually defined by the resistance to lateral deflection or bowing. When the column is a long column (elastic failure), the critical load is related by the Euler equation:

$$P_{crit} = \frac{C\pi^2 EI}{L^2}$$

Where P_{crit} = buckling load [lbs]
E = modulus of elasticity [psi]
I = least moment of inertia of column cross–section [in⁴]
L = column length [in]
C = end fixity coefficient

From this relationship, the properties of the column (E, I, L, and C) determine the maximum load carrying strength. The end fixity coefficient C defines the restraint given the ends of the column. If the ends of the column are pinned in place so they are free to rotate but cannot translate, the end fixity coefficient is 1.0. If both ends are fixed in place and rigidly restrained from rotation as well as translation, the end fixity coefficient is 4.0. The case of full fixity of both column ends is rare because of the flexibility and elastic nature of the supporting structure and values for C rarely exceed 2.5 to 3.0 with typical structures. \

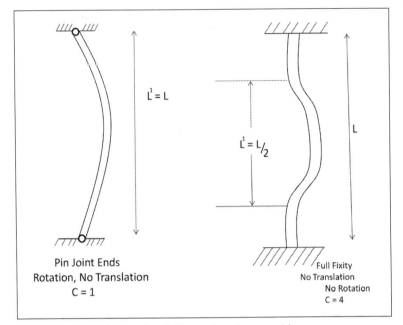

Figure 7.2 Effect of end constraints

To account for various conditions of restraint, an equivalent pin-ended length of the column L' is defined:

$$L' = \frac{L}{\sqrt{C}}$$

where L' = equivalent pin-ended length of column
L = actual column length
C = end fixity coefficient

Thus, a column which has full fixity at both ends (C = 4.0) would have an equivalent pin-ended length which is one-half the actual length of the column. Figure 7.2 illustrates this case. Figure 7.2a provides more details concerning effects of different boundary conditions on critical buckling load..

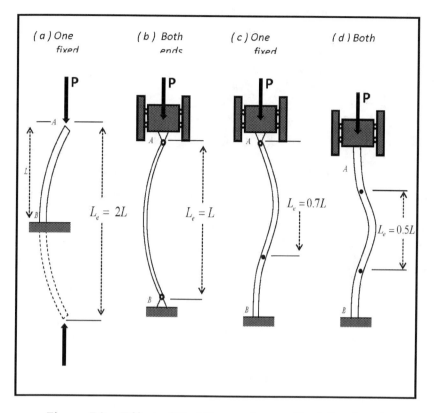

Figure 7.2a: Effect of End Constraints on Buckling Load

The effect of column length and cross-section on the critical load can be combined by a certain modification of the Euler equation and the definition of the radius of gyration (ρ)

$$\rho = \sqrt{\frac{I}{A}}$$

$$P_{crit} = \frac{C\pi^2 EI}{L^2} = \frac{\pi^2 EI}{(L')^2}$$

$$\frac{P_{crit}}{A} = \frac{\pi^2 EI}{(L')^2 A'} \quad \sigma_{crit} = \frac{\pi^2 E}{(L'/\rho)^2} = F_c$$

Where σ_{crit} = allowable compressive stress [psi] = F_c

E = modulus of elasticity [psi]

L' = equivalent pin-ended length of column [in] = $L / \sqrt{c}$

ρ = radius of gyration [in]

(L' / ρ) = slenderness ratio

By the use of the equation for σ_{crit}, the allowable stress of the long column can be determined and illustrated as in Figure 7.3.

The behavior of the elastic long column shows the principal effect of the slenderness ratio for the long column (e.g., if the slenderness ratio is doubled, the allowable column stress is reduced to one fourth the original value). Actually, the range of allowable column stress for which elastic column behavior applies is relatively small. The highest column stress which can be predicated on the Euler equation is approximately 40 to 50% of the upper limit of column stress at small slenderness ratios.

The *transitional slenderness ratio* defines the boundary between elastic and inelastic column behavior. The values of (L'/ρ) greater than their transition boundary define the regime of the elastic long column. Values of (L'/ρ) less than this transition boundary define the regime of the inelastic short column.

Since the aircraft and missile structure must be efficient and light, the components must operate at the highest practical stress to reduce structural weight. Since the long column has relatively low values of allowable stress, the structural efficiency is quite low and the long column is rarely encountered in the primary load-carrying structure of any aircraft or missile. In order to achieve high allowable column stress and produce an efficient structure, the compression members of the primary structure are inevitably of the short column type.

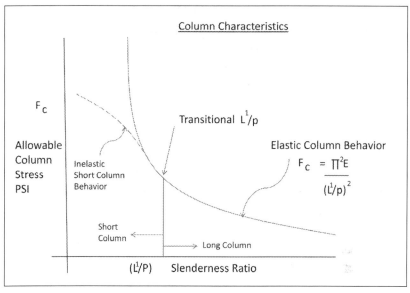

Figure 7.3 Column characteristics

The behavior of the short column can be explained in part by the decrease in stiffness at stresses above the proportional limit. The tangent modulus (E_t) is the instantaneous slope of the stress-strain diagram and this quantity portrays the instantaneous stiffness of the material at a given stress. The secant modulus (E_s) is the proportion of applied stress and resulting strain and this quantity portrays a cumulative stiffness at a particular level of stress. Both the tangent and secant modulus can be used to approximate the allowable stress of short columns when the primary mode of failure is a bowing or buckling instability.

$$\sigma_c = \frac{\pi^2 E_t}{\left(L'/\rho\right)^2} \text{ or alternatively } \sigma_c = \frac{\pi^2 E_s}{\left(L'/\rho\right)^2}$$

The use of the tangent modulus (E_t) will usually predict conservative values for the allowable column stress while use of the secant modulus (E_s) will usually predict optimistic values for the allowable column stress. While the application of these equations is helpful in establishing an appreciation for short column behavior, the results of these equations generally are not of sufficient accuracy for use in detail design and more exact data must be derived. The combination of comprehensive buckling theory and extensive component tests provide the column data for use in structural design. The typical aircraft or missile structure is composed of relatively lightweight elements and the buckling considerations are of great importance in their structural design.

Since lightweight structures necessitate high stress, the majority of compression structures are of the short column type. To achieve a high allowable stress, the slenderness ratio must be small and the cross-section must be distributed to increase the radius of gyration. As a result, the thin elements of the cross-section may develop a local instability before primary failure occurs. When local instability produces a distortion of the cross-section prior to complete failure, the mode of failure is termed *local crippling*. This sort of local failure is concerned with sections composed of thin sheet material and can be related directly to the buckling of panels and thin sheet sections.

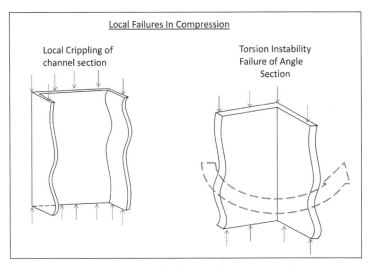

Figure 7.4 Local failures in compression

Another problem encountered in compression structures is *torsional instability*. Certain cross-section shapes of columns are made up of thin sheet metal which do not form completely closed shapes. Simple channel and angle sections form open sections with relatively low torsional stiffness and these sections allow a mode of compression failure which is combined bending-torsion instability. As the column is gradually loaded with compression, slight but noticeable bending deflections occur prior to reaching the critical load. The variation of bending moment along the length of the column gives evidence to the existence of a lateral shear force throughout the length of the column. When this lateral force is eccentric to the elastic center of the cross-section, a twisting moment exists along the longitudinal axis of the column. If the section is relatively flexible in torsion, the initial failure may be a twisting instability. Figure 7.4 illustrates the mode of failure typical with torsional instability in a simple angle section.

The buckling strength of thin plates or sheet sections is of considerable importance in aircraft construction because of the predominance of thin sheet metal elements. If a rectangular panel is subjected to edgewise compression loading, the allowable stress will be defined by buckling instability of the panel. Figure 7.5 illustrates the case of a rectangular panel of thickness t subjected to a compressive stress along the panel edge of dimension b. Suppose that all edges are pinned or guided such that the edges are free to rotate but are restrained against translation.

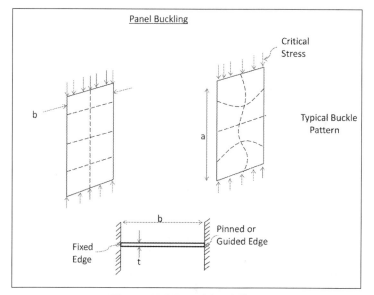

Figure 7.5 Panel buckling

As the compressive edge stress is increased to some critical value, the panel will deflect into a buckled wave form typical of Figure 7.5. The exact wave form into which the panel will buckle will be determined by the aspect ratio of the panel (a/b) and the type of restraint furnished by the various edges. In each case, the wave pattern will represent a minimum energy of the deflected or buckled panel.

The critical compressive stress of the rectangular panel subjected to edgewise compression can be expressed by the following equation:

$$\sigma_{crit} = KE\left(t/b\right)^2$$

Where σ_{crit} = Panel buckling stress [psi]
               K = fixity coefficient
               E = modulus of elasticity [psi]
               t = plate thickness [in]
               b   = length of the loaded edge [in]

This relationship defines the powerful effect of the major dimension of the panel. For example, if the thickness of a given panel is doubled, the critical stress is increased by a factor of four and the critical load is increased by a factor of eight. The powerful effect of thickness is favorable to the use of lightweight metals to provide high proportions between buckling strength and weight.

The fixity coefficient (K) of the panel buckling equation is dependent upon the panel aspect ratio (a/b) and the type of restraint furnished by the edges of the plate. If the panel aspect ratio is relatively large, the aspect ratio has no significant effect and the edge restraint is the factor of greatest influence. Typical values of K are listed in Table 7.1:

Table 7.1 Typical values for the fixity coefficient K

K	Edge restraint conditions
6.3	All four edges fixed
3.6	All four edges pinned or guided
1.2	Loaded edges guided, one unloaded edge fixed and one free
0.46	Loaded edges guided, one unloaded edge guided and one free

Actually, the restraint furnished by the loaded edges is not of great importance when the panel aspect ratio is large. If the panel aspect ratio is much greater than two, the half-wave length of the buckles is about the width of the panel regardless of the restraint furnished by the loaded edges. Since the buckling stress is a characteristic of the half-wave length, no difference in buckling stress will exist if the half-wave lengths are identical. For example, at high values of (a/b), the fixity coefficient for loaded edges guided and unloaded edges fixed is 6.3. This is equal to the fixity coefficient when at high aspect ratios and all four edges are fixed.

The behavior of compression panels in the elastic range is accurately described by the previous equation. The behavior of compression panels in the inelastic range is difficult to describe accurately and the panel behavior is likened to column behavior in the following manner:

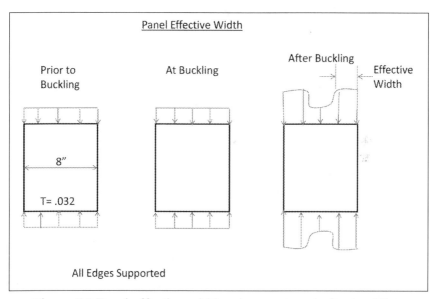

Figure 7.6 Panel effective width prior to, at, and after buckling

The critical stress for a column is: $F_c = \dfrac{\pi^2 E}{\left(L^1 / \rho\right)^2}$

The critical stress for a panel is: $F_c = \sigma_{crit} = KE\left(t / b\right)^2$

If it is assumed that the properties of the material are essentially the same for the panel and the column, a relationship between the panel and the column is established:

$$\frac{L'}{\rho} = \frac{\pi}{\sqrt{K}}\left(\frac{b}{t}\right)$$

Thus, the column curves developed for a particular alloy can be used to predict the strength of panel members of the same material.

The ultimate strength of supported panel members will be higher than the stress which causes buckling. Since the portions of the panel adjacent to the supports are restrained and are relatively stiffer, these areas will continue to take higher stress after buckling of the control areas takes place. Figure 7.6 illustrates the change in stress distribution through a panel during the process of buckling. To account for the portions of the panel which operate at a stress greater than the buckling stress, an *effective width* of panel is considered to operate at the edge stress. While there are fairly accurate means of determining the panel stress distribution and the effective width existing at the edges, most preliminary estimates of panel strength assumes this effective width to be approximately fifteen times the panel thickness (15t).

As an example, consider the plate of Figure 7.6 to be of aluminum alloy 0.032 inches thick and 8 inches wide along the loaded edge. If the panel has all edges guided, the buckling stress of the panel would be:

$$\sigma_{crit} = KE\left(t/b\right)^2 = \left(3.6\right)\left(10.5\times10^6\, psi\right)\left(\frac{0.032in}{8in}\right)^2 = 604.8\, psi$$

This stress corresponds to a panel buckling load of:

$$P_{crit} = \sigma_{crit}tb = \left(604.8\ psi\right)\left(0.032\ in\right)\left(8\ in\right) = 154.8\ lbs$$

The edge stress at maximum panel load usually approaches, but rarely exceeds, the compression yield stress σ_{cy}. If the effective widths at each edge of the panel are assumed to be 15t, the panel would be capable of an ultimate load of:

$$P = üüüöüüeüüßüü)t\sigma_{cy} = (\quad)(\quad in)^2(\quad psi) = \quad lbs$$

This difference between the panel ultimate load and the panel buckling load points to the favorable effect obtained by the edge restraint, which allows portions of the panel to develop much higher stress than the panel buckling stress. For example, consider a 0.032 inch thick aluminum alloy plate with an edge width of 4 inches and a length of 10 inches. If this flat plate is unsupported, the failing load of the part is approximately 12 lbs.

As shown in Figure 7.7, a longitudinal bend of this sheet to form an angle section will raise the maximum load for local failure to 150 lbs. If the panel is formed into a Z-section of equal element lengths, greater restraint is furnished to the elements and more effective widths come into effect. The maximum load for local failure of the Z-section would be approximately 300 lbs.

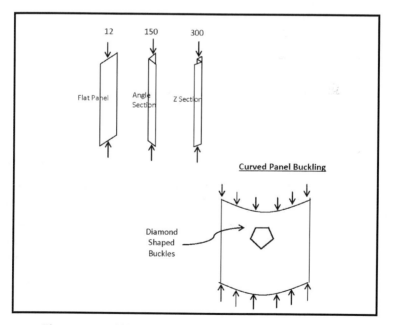

Figure 7.7 Different buckling loads for local failure

The compression buckling of curved panels is strongly affected by the thickness and curvature of the sheet. The mode of failure of a curved panel subjected to edgewise compression is a buckle pattern of diamond shaped indentations in the sheet. The elastic buckling stress of such a curved panel can be determined from the following equation:

$$\sigma_{crit} = 0.3E\left(t/r\right)$$

Where t = sheet thickness [in]
        r = radius of curvature [in]

A curved panel of a pressurized vessel is capable of sustaining higher buckling stress from the favorable restraint furnished by the internal pressure.

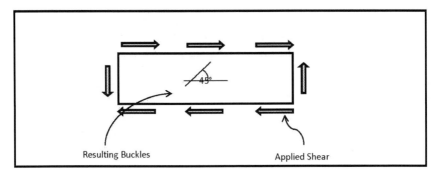

Figure 7.8 Shear buckling of a thin panel

Panel elements subjected to shear stress may buckle and this instability could be due to the component stress of compression resulting from applied shear. Figure 7.8 illustrates the mode of shear buckling of a thin, rectangular panel supported on all edges. The component compression stress exists as a maximum at 45° and the principal wave direction of the buckles will be along this direction. The elastic buckling stress of a shear panel of high aspect ratio can be predicted by an equation similar to that used for compression panels.

$$\tau_{crit} = KE(t/b)^2$$

Where τ_{crit} = shear stress to cause panel buckling [psi]
             K = fixity coefficient
             E = modulus of elasticity [psi]
             t = plate thickness [in]
             b = length of the loaded edge [in]

When the panel has all four edges guided, the fixity coefficient K is 5.0 and when all four edges are fixed, the fixity coefficient is 8.0. Ordinarily, full fixity is not attained practically in a typical structure since the restraining elements provide only elastic support. However, the panel dimensions can be reduced by the addition of stiffening elements and the buckling stress can be increased by this means.

Most shear panels can be classified as one of two general types. The *shear resistant panel* is designed to prevent buckling. In this case, buckling would constitute an ultimate load condition. A *tension field shear panel* allows the panel to resist shear loads greater than the initial buckling load by the development of large tension stresses along the folds of the buckles. The tension field shear panel requires considerable support to prevent collapse of the panel. The large diagonal tension existing in the panel causes compression loads in the frame surrounding the panel and places the stiffeners in compression. The tension field shear panel is widely used in the construction of relatively deep beams since it provides a minimum weight structure. The mode of failure of the tension field shear panel is generally a tensile failure of the sheet due to the action of the large diagonal tension stresses. For the particular case of a very thin tension field panel, the diagonal tension stress is twice the shear stress applied to the edges of the panel.

The combined stability problem of a panel supported by stringers or stiffeners is quite complex. Obviously, a close spacing of stringers or stiffeners along a panel will reduce the dimensions of the panel and provide relatively higher operating stresses for the panel. However, more numerous stiffening elements add additional weight, which is an important consideration. The optimum structure will be some arrangement of skin and stiffener which produces the minimum weight structure. In order to provide effective restraint to the edges of the panel, the stiffening elements should be quite rigid in torsion and should have relatively high allowable stresses. Thus, a closed box or hat section would provide greater panel restraint than an open angle or Z-section. Figure 7.9 illustrates typical panel stiffeners.

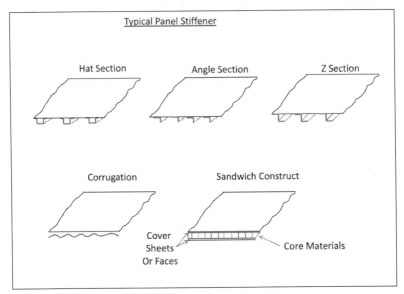

Figure 7.9 Typical panel stiffeners

A corrugated skin may be attached to a curved or flat panel to stiffen the outer panel and increase the buckling strength. Various corrugated forms can be riveted or welded to the panel at the points of contact. Sandwich panels provide a means of increasing the resistance to buckling and improving the strength-to-weight ratios of panel members. The two cover sheets or panel faces are separated by a lightweight core of foam, balsa wood, or plastic fiber and metal honeycomb material. The cementing or bonding of the sheets and the core create a panel structure which has high stiffness and high resistance to buckling.

Buckling of the exterior surfaces of an aircraft or missile may or may not be detrimental. Generally, it is necessary to maintain high surface smoothness for an aircraft exterior when at the cruise condition. If buckling of the exterior does occur in accelerated flight, the change in contours should be held to a minimum and no adverse aerodynamic effects should take place. Of course, the high speed aircraft or missile demands more exact surface contours and will require relatively stiffer surfaces. Low speed aircraft can tolerate somewhat less exact contours if the leading edges of all surfaces remain quite smooth and exact. Of course, the missile or space craft in non-atmospheric flight has no particular demand for smooth exterior surfaces.

Initial curvature or eccentricity of loading can have a considerable effect on the critical load of a short column or any inelastic compression structure. Actually, small amounts of eccentricity and initial curvature do not change the critical load of the purely elastic long column. If the stresses remain in the elastic range, there is no effect on the critical load of the column. Figure 7.10 illustrates this fact by the variation of end deflection with applied compression load P. For the case of the long elastic column, increasing eccentricity increases the end deflection upon application of load, but the maximum value of end load which the column can sustain is unchanged.

The case of the short column is quite different since an eccentric loading or initial curvature accentuates the inelastic behavior of the structure and reduces the critical load. As shown in Figure 7.10, increasing eccentricity increases the end deflection with application of load and decreases the maximum value of end load which the column can sustain.

Since the majority of all primary compression structures are of the short column type, initial eccentricity and curvature must be held to a minimum during design and manufacture. Also, any damage to a primary compression structure which might occur during operation must be carefully evaluated for its effect on compression ultimate strength of the structure. Since buckling of primary structure constitutes an ultimate load condition, considerable damage could reduce the buckling loads of certain structures to some value near the limit load condition.

When a column is subject to lateral loads in addition to the axial compression load, the critical compression load is considerably reduced. Such a beam column structure is encountered quite often in typical structures, but it requires special consideration. The interaction of the beam and column loading is most critical and difficult to predict accurately when the resulting stresses produce an inelastic behavior.

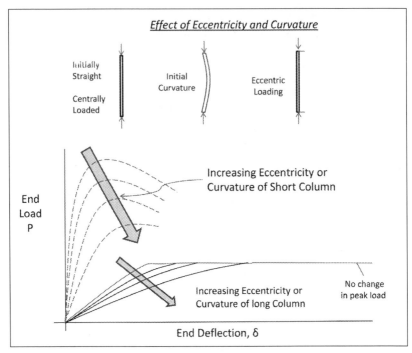

Figure 7.10: Effect of eccentricity and curvature

When high temperatures exist in the structure, creep considerations provide an additional complication to the analysis of the stability of compression structures. Creep buckling is possible in that any slight initial bowing of the column under load will increase with time. The increase in material strain with time due to creep will allow the column to fail at some stress less than the simple static column. The effect is to produce column behavior in which the critical load is dependent upon time as well as initial curvature. In addition, high temperatures reduce the inherent stiffness of the metal and lower the allowable column stresses. High speed, low altitude flight is most likely to generate most critical temperatures.

Summary: Structural Instability called "Buckling"

This chapter defines buckling, states the fundamental requirement for buckling; discusses the influence of boundary conditions on the buckling problem, and the importance of changes in boundary conditions; and discusses the **load** vs. deflection behavior of columns, plates and shells, including (especially!) their post-buckling behavior.

Definition: Buckling is a process occurring at or above a "critical buckling load," in which relatively large, non-proportional changes in stresses and deflections result from small changes in load or in the effective point of load application. The fundamental requirement for buckling is compressive stress. Buckling is considered as instability. Figure 7.11 shows stability concepts.

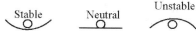

Figure 7.11 Stability visual concepts

Column Buckling: Figure 7.12 shows Euler column formula and load.

Figure 7.12 Euler Column Load

Other boundary conditions

Euler columns have hinges at both ends. But what if we looked for the deflected shape of the Euler column in the deflected shape of other columns, and then used the fraction of the total column length, which has that shape as the "effective length" of that column (refer Figure 7.2a). L'= "effective length"

"Fixed-Fixed" boundary conditions: Mathematics shows a factor of 16 changes in pcr for idealized boundary conditions. For the Fixed-Fixed Condition the critical load is 4 times that of Fixed-Pin ended B.C. (refer Figure 7.2 and corresponding equation).

"Fixed-Free" boundary conditions; $P_{cr} = \dfrac{EI\Pi^2}{4L^2}$

119

Changing the boundary conditions has an enormous effect on the buckling load! How are boundary conditions changed?

A) Fasteners unfastened create a different boundary condition. Dr. Frank gives an example of his VW van which he used to use to transport kids for sports trips. Sometimes the kids would fail to completely close the door, thus leaving it partially open. Driving on uneven country roads for long and repeated times led to the van door frames becoming elongated, thus making it impossible to properly close the doors. Driving the van with improper closure of the doors created different boundary conditions which led to undesirable deformation of the van door assembly.

B) Improper construction using lower quality material can cause undesirable accidents. For an example, in 1978 the **Hartford Civic Center Coliseum** collapsed in the early morning hours under the weight of snow during the largest snow storm in the five year life span of the **arena**. http://failures.wikispaces.com/Hartford+Civic+Center+(Johnson)

C) Design can be changed by Production engineers? Early days of manufacturing aircrafts, technicians would use fasteners without paying attention to exact locations in the structure. A situation arose whereby aircraft required replacement of certain parts. Since it was not specified that fasteners be attached at exact locations, the customer spent more money than required for making the aircraft usable. Military standard 882 was a result of such actions by manufacturers.

D) Change in point of load application may cause undesirable deformation. The Mercury Comet automobile was designed such that towing the vehicle could cause the frame to buckle.

Load vs Deflection Histories: What do structural elements do after they've buckled? A column will continue to carry its buckling load, but it will not accept more load after it has buckled. Load vs. deflection diagram is given in figure 7.1.

Plate Behavior:

If a rectangular panel is subjected to edgewise compression loading, the allowable stress will be defined by buckling instability of the panel. Figure 7.5 illustrates the case of a rectangular panel of thickness t subjected to a compressive stress along the panel edge of dimension b. Suppose that all edges are pinned or guided such that the edges are free to rotate but are restrained against translation. As the compressive edge stress is increased to some critical value, the panel will deflect into a buckled wave form typical of Figure 7.5. The exact wave form into which the panel will buckle will be determined by the aspect ratio of the panel (a/b) and the type of restraint furnished by the various edges. The critical compressive stress of the rectangular panel subjected to edgewise compression can be expressed by the following equation:

$$\sigma_{crit} = KE(t/b)^2$$

Where
σ_{crit} = panel buckling stress [psi]
K = fixity coefficient
E = modulus of elasticity [psi]
t = plate thickness [in]
b = length of the loaded edge [in]

The fixity coefficient (K) of the panel buckling equation is dependent upon the panel aspect ratio (a/b) and the type of restraint furnished by the edges of the plate. If the panel aspect ratio is relatively large, the aspect ratio has no significant effect and the edge restraint is the factor of greatest influence. Typical values of K are listed in Table 7.1

Shell Behavior:

Buckle shapes tell you what kind of load caused shell buckling. Axial compressive load causes shell failure. The following example (Figure 7.13) demonstrates a simple case of axial compressive load and transverse wall load on a coke can causing buckling failure. When shell buckles, it sheds its load and is unable to carry the load any more.

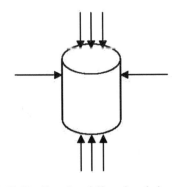

Figure 7.13 Coke Can buckling load demonstration.

Thus, axial, torsional and external pressure may cause buckling in shells. An expert from your depot can read the buckled pattern for you and tell you what kind of load caused the buckling.

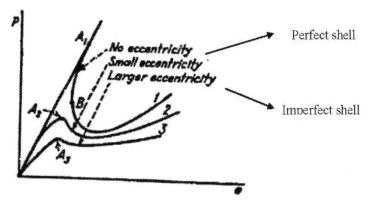

Figure 7.14 Load vs. deflection diagram

Google search on "Euler Beam Bending" provides more description on Buckling of beams, plates and shells.

8

STRESS CONCENTRATION

All mechanical parts have certain shapes, variations of cross-section, and discontinuities which create an undesirable magnification of local stresses. Of course, such discontinuities must be held to a minimum because of the possible reduction of strength.

The effect of a physical discontinuity (or stress concentration) is to alter the distribution of stress on the cross-section and cause a peak stress at the discontinuity. This peak stress may be considerably greater than the ordinary stress distributed on the same cross-section without the discontinuity. The discontinuity creating the magnification may take many different forms (e.g., sudden change of shaft diameter, notches, grooves, keyways, rivet holes, corrosion pits, tool marks, etc.). In addition, it must be considered that all ordinary materials are not perfectly homogeneous and there is the possibility of minute flaws, fissures, and microscopic cracks in typical metals.

The manner in which discontinuities create local magnification of stress distribution is analogous to the manner in which objects placed in a fluid stream cause local magnification of flow velocity. In fact, there is an exact analogy between pressure distribution in incompressible, perfect fluid flow and stress distribution in perfectly elastic structures. If a cylinder were placed in a fluid stream, there would be flow stagnation at the leading and trailing edges, causing an increase in static pressure at these points equal to the free stream dynamic pressure. As the flow proceeds around the cylinder, a maximum velocity is created at the extremities which is twice the free stream velocity. This local magnification of velocity would create a drop in pressure (or suction) which is three times the free stream dynamic pressure. Such a situation is illustrated in Figure 8.1 by the variation of pressure coefficients around the surface of the cylinder in fluid flow. At the leading edge (point A), $C_p = +1$ and at the extremity (point B), $C_p = -3$.

In the analogous condition, suppose a circular hole exists in an infinitely wide elastic plate. When this plate is subjected to a uniform axial stress, the local stress at the hole is distributed in a fashion identical to the pressure on the cylinder. If a uniform tension stress of 1,000 psi is applied to the plate at point A, the stress is a compression stress of 1,000 psi. At point B, the stress has changed to tension and is magnified to a value of 3,000 psi.

This fundamental analogy is quite important since contours which would create high local velocities and pressures in a fluid stream would cause the same high local magnification of stress in a structure. Significant changes in flow, pressure, and velocity are relatively localized and occur at the surface of an object in a flow stream. In the same sense, significant changes in stress at a stress concentration are highly localized and exist as peak values at the surface of the discontinuity. The high local stresses decrease rapidly with increasing distance from the discontinuity and rapidly approach the normal stress a short distance away. A fact common to any local stress concentration is that the critical magnification of stress is confined to the immediate area of the discontinuity.

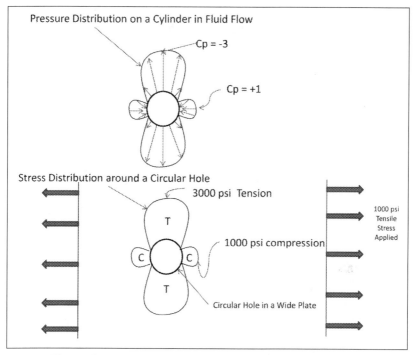

Figure 8.1 Pressure distribution on a cylinder in fluid flow and stress distribution around a circular hole

To describe the effect of a discontinuity or stress concentration, it is appropriate to consider the degree of stress magnification which takes place. This consideration is accomplished by definition of the stress concentration factor K:

$$K = \frac{\text{maximum stress at discontinuity}}{\text{average stress on net cross-section}}$$

or as may be more applicable in certain conditions,

$$K = \frac{\text{maximum stress at discontinuity}}{\text{stress calculated elementary theory}}$$

By these definitions, the circular hole in the wide plate subjected to axial stress would create a stress concentration factor of three (i.e., K=3.0).

The elementary equations for the maximum stresses due to axial bending and torsion loads remain applicable. However, in the case of a stress concentration, these maximum stresses are obtained by including the effect of the discontinuity. Figure 8.2 illustrates these applications.

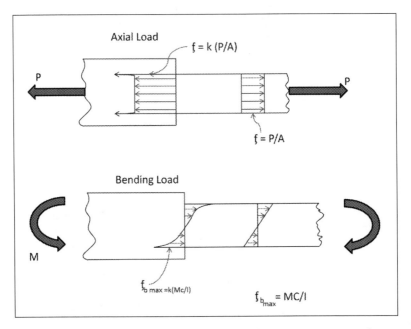

Figure 8.2 Axial and bending loads with stress concentration.

Axial or shear loads : $\sigma \dfrac{P}{A}$ *becomes* $\sigma = K\left(\dfrac{P}{A}\right)$

Bending loads : $\sigma_{b\,\max} \dfrac{Mc}{I}$ *becomes* $\sigma_{b\,\max} = K\left(\dfrac{Mc}{I}\right)$

Torsion loads : $\tau_{\max} \dfrac{Tc}{J}$ *becomes* $\tau_{\max} = K\left(\dfrac{Tc}{J}\right)$

It is important to distinguish between the stress concentration factors for shear and that for axial stress. For example, a circular hole in an infinite plate subject to a single axial stress causes a stress concentration factor of 3.0. On the other hand, the same circular hole in an infinite plate subject to pure shear causes a stress concentration factor of 4.0. Since a pure shear stress could be made equivalent to equal stresses of tension and compression acting on lines 90° apart, the stress concentration factor is K=3+1=4.

It is implied by the previous definitions of the stress concentration factor that the degree of magnification is determined by the relative shape and contour of the discontinuity. In other words, the stress concentration factor is determined by relative shape rather than physical size. For this reason, a relatively small but severe discontinuity such as a scratch, tool mark, corrosion pit, etc. may create a relatively high value of K. A circular hole in a very wide plate subject to axial stress causes K=3.0. Within practical limits, K=3.0 for all sizes of circular holes.

An important application of the previous statements is during the repair of damaged structures. Suppose that a turbine component has an obvious stress concentration due to impact with foreign objects. If an attempt is made to file out this large discontinuity, the resulting small file marks may create a greater degree of stress concentration than the original discontinuity. Great care must be taken to insure a fine, smooth surface after repair work. Fine grinding is usually an accepted method which is preferable to filing and then polishing.

In order to appreciate the principal factors affecting the degree of stress concentration, consider an elliptical hole in an infinite plate subject to axial stress. Figure 8.3 illustrates such a stress concentration which has principal dimensions a and b; and an aspect ratio of a/b. The specific equation which defines the ellipse is:

$$\frac{x^2}{a^2} + \frac{y^2}{b^2} = 1$$

Thus, the ellipse of dimensions a and b define a specific contour and shape to the discontinuity in the plate. From the theory of elasticity, the stress concentration factor for the ellipse is defined as

$$K = 1 + 2\left(\frac{a}{b}\right)$$

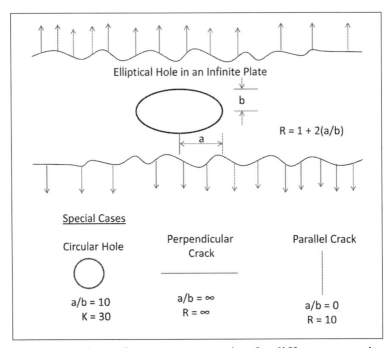

Figure 8.3 Values of stress concentration for different scenarios.

Hence, the stress concentration factor for the ellipse is defined as a function of the aspect ratio a/b. Some special cases of the ellipse are of significance in appreciating the effect of shape or the magnitude of the stress concentration factor.

(1) For a circle hole: $a = b$ or $\dfrac{a}{b} = 1$ $K = 1 + 2(1) = 3$

This is identical to the factor originally defined by the hydrodynamic analogy.

(2) For a crack perpendicular to applied axial stress: $b = 0$ or $\dfrac{a}{b} = \infty$ $K = 1 + 2(\infty) = \infty$

The significance of this result is important since an infinite value for K implies that such a crack causes the most severe possible magnification of stress. Of course, actual stresses do not reach infinite values since there are specific limitations to the application of the original equation. Upon the application of stress, the crack would open and a/b < ∞. Also, local yielding and plastic flow would not allow infinite stress at the edges of the crack. However, these statements should not reduce the impression of the severe nature of such a stress concentration. It is significant that such cracks will continue propagating when in the presence of cyclic tensile stress.

(3) For a crack parallel to applied axial stress: $a = 0$ or $\dfrac{a}{b} = 0$ $K = 1 + 2(0) = 1$

This implies that such a discontinuity is not significant for the particular direction of applied stress since the stress is not increased above the average stress in the plate. Thus, there is a preferred orientation of stress concentrations parallel to the applied axial stress. The application of this idea is important in the grinding of surfaces during surface repairs. The grinding (or filing) direction should be along the direction of anticipated principal tensile stress. Also, there is explanation for the behavior of most materials subject to cross grain loading. Generally, most metals demonstrate superior strength characteristics when the principal stress direction is along the apparent grain direction.

In addition to the theory of elasticity and the hydrodynamic analogies, stress concentration factors can be defined by photoelastic methods. In this case, acrylic plastic models subjected to particular loadings will illustrate stress distribution with polarized light. Also, structural models may be coated with a brittle lacquer which will illustrate the distribution of critical levels of stress. The principal advantage of the photoelastic and brittle lacquer techniques is that highly complex structural shapes and loadings may be investigated.

A particularly difficult situation exists when stress concentrations are superimposed. As shown in Figure 8.4(a), a circular hole in a wide plate causes K=3.0. A sharp V-notch along the edge of a wide plate causes K=6.0.

If a circular hole is drilled in a wide plate and poor technique and poor quality tools are used, burring or galling of the hole surface may result. Suppose that the poor drill or bore surface causes the equivalent of a sharp v-notch. In this case, the plate stress is magnified by a factor of three by the existence of the circular hole. Then, this peak stress is again magnified by a factor of six because of the sharp notch. With this superimposition of stress concentrations, the net magnification of stress is K=(3)(6)=18.

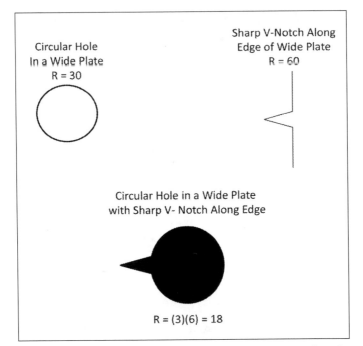

Figure 8.4 Circular hole and v-notch stress concentration factors

Poor quality control, poor shop practices, and poor or inadequate tools can lead to structures with severe and damaging stress concentrations. Such conditions could provoke cracks in structures which are difficult to detect by ordinary inspection methods.

The effect of stress concentrations on the strength of structures depends to a great extent on the ductility of the material and the manner of loading. In fact, it is quite necessary to make a clear distinction between the effect of ductility and the effect of the type of loading. The essential effects are summarized in the following discussion.

8.1 EFFECT ON STATIC STRENGTH: BRITTLE MATERIALS

If a material has no significant ductility and approaches the behavior of glass, ceramics, carbides, etc., the effect of stress concentrations is severe and damaging. Of course, this domain would include normally ductile metals which are subject to extremely low temperatures and suffer considerable loss of ductility. In these cases, any stress concentration would cause the critical stress to be reached at considerably less than the normal failing load. The failing load of a brittle cross-section would vary inversely with the stress concentration factor K. For example, a structure of brittle material which has a circular hole (K=3) within the section would have a strength of only one-third that of the same section without the stress concentration.

The notching of a glass plate with a diamond or carbide point localizes the fracture and requires very little load to cause failure. The obvious conclusion is that very high strength brittle elements are extremely sensitive to stress concentrations. Such structures must have very exact contours since only the most mild stress concentrations can be tolerated.

8.2 EFFECT ON STATIC STRENGTH: DUCTILE MATERIALS

If a material has very high ductility, it can be quite tolerant of the most severe stress concentrations. The advantage of the ductile nature of the material is that the high local stresses at discontinuities will simply cause premature local plastic flow and yielding. In this manner, the stress at the discontinuity is redistributed and no truly adverse effect is encountered. Figure 8.5 depicts the stress-strain diagram along with a representative distribution of stress around a circular hole in a highly ductile material. The material is assumed to have a yield strength of 30,000 psi and an ultimate strength of 60,000 psi with extremely large elongation.

The circular hole provides K=3 when the behavior is elastic. Thus, an applied stress of 10,000 psi creates a peak stress of 30,000 psi at the edge of the hole, just on the verge of yielding. However, an applied stress of 20,000 psi does not create a peak stress of 60,000 psi at the hole since considerable plastic deformation would occur as the local stress exceeded 30,000 psi. Initially, the strain may be magnified by a factor of three, but causes stresses and flow into the plastic range. With subsequent increase in applied stress, the stress distribution becomes more uniform and as ultimate stress is approached, only a small degree of magnification will exist at the hole. To be sure, initial fracture is most likely to begin at the hole, but the stress distribution on the cross-section is nearly uniform.

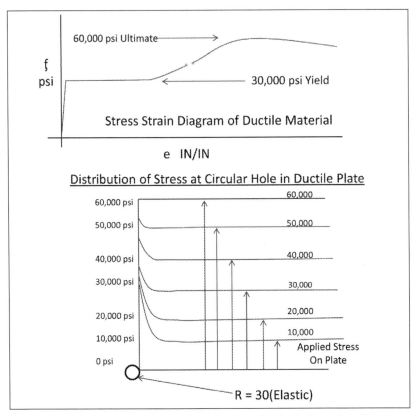

Figure 8.5 Distribution of stress at a circular hole in a ductile plate

The behavior typical of the ductile material illustrates the fundamental requirement of ductility in ordinary structures. If sufficient ductility is available, the usual stress concentrations can be tolerated without significant reduction in static ultimate strength. The large plastic deformations in the vicinity of the discontinuities allow a more uniform redistribution of stress. However, the enigma is that materials which possess high intrinsic strength do so by virtue of limiting plastic deformation. The very high strength ceramics, cermets, carbides, etc. then require great care and precision in manufacture and handling.

8.3 EFFECT ON RESISTANCE TO ENERGY OR IMPACT LOADS

The principal characteristic of the impact load is that the load is applied and developed in a very short period of time. The energy input into the structure must be stored or absorbed without ill effect. If stress concentrations exist in a part subjected to impact stress, plastic flow and yielding may be limited because there is not sufficient time available for that to take place. As a result, the introduction of stress concentrations will increase the sensitivity of a part to impact or energy type loads.

Generally, less ductile materials will be more severely affected by energy or impact loads in that there will be an inherent brittle behavior. In addition, normally ductile materials operating at very low temperatures will be more severely affected by impact or energy loads, especially with stress concentrations present.

8.4 EFFECT ON FATIGUE STRENGTH

Metal fatigue is the result of some critical accumulation of cyclic plastic strain. When a part is subjected to repeated or cyclic stress, fine slip is generated within the crystalline structure. As this slip deteriorates, microscopic cracks form and become more numerous with continued cyclic stress. A critical crack is formed and propagates into the cross-section until a final rupture occurs in the remaining area. Underlying this entire sequence of fatigue failure is the generation of cyclic plastic strain. Thus, any increase in stress on a material will increase the magnitude of plastic strain and allow fatigue failure to occur with fewer applications of stress. Since discontinuities will provide local increase in stress or strain, the action of stress concentrations will cause a serious reduction in resistance to repeated or cyclic loads. It is a well-established fact that all fatigue failures which occur in service show an origin or beginning crack located at a distinguishable stress concentration.

In order for a part to provide satisfactory service life, there must be sufficient resistance to fatigue. A major part of this fatigue resistance is determined by the number and severity of stress concentrations existing in the part. For this reason, it is imperative that primary structural components be manufactured of the highest quality materials and the basic design features must allow only the most mild stress concentrations. The maintenance and care given these parts in service must insure that no additional stress concentrations are incurred because of poor practices. Any maintenance or repair of structures must provide adequate edge distance for holes, satisfactory surface finish, etc. Tool marks, scratches, file marks, poorly drilled holes, etc. may contribute to a serious reduction in service life.

The effect of material characteristics in this instance is somewhat similar to the situation for static loads. The brittle materials are generally of higher intrinsic strength and thus, a higher operating stress is tolerated. However, such materials are inherently notch-sensitive and stress concentrations cause considerable change in fatigue resistance. The more ductile materials are less sensitive to stress concentrations, but their tolerance of allowable operating stresses is relatively low. The favorable effect of the more ductile material is that random fatigue loading is likely to produce initial yielding in the vicinity of stress concentrations. If favorable residual compression stresses are created, there is a significant resistance to crack formation and crack propagation.

The ductile quality of the material does have a significant effect on the resistance to crack formation and crack propagation. This feature is so important that a long life structure which must accommodate fairly typical stress concentrations may generate a preference for a lower strength, but less notch-sensitive material. A similar alloy of higher static strength may be so sensitive to stress concentrations that the typical fatigue resistance is inferior.

The most significant effects of stress concentrations occur with the following conditions:

(1) Low ductility and toughness

(2) Impact or energy type loads

(3) Repeated or cyclic stressing

In addition, operation of the structure at very low temperatures will enhance the vulnerability to stress concentrations.

8.5 Commonly Used Terms

Definition: A Stress Concentration (Stress Riser) is a condition in which high localized stresses are produced as a result of an abrupt change in geometry and/or material properties. Mathematicians use a better word:

"discontinuity and/or singularity" instead of abrupt change in geometry and change in material properties.

Examples of macroscopic stress risers:

1. Holes and cutouts can reduce life by as much as 50%,
2. Tool marks: In 1892 Sonderick (G.B.) showed that a mere 0.003" deep scratch reduces life of steel specimens by 40%. In 1962 par teams refinish F8 wing fold hinges to remove scratches. Early '70s – A6's were grounded because tool marks in the wing attachment fittings were causing failures.

In 1999, H-60's were grounded for 30 hour inspection of the main rotor damper. Mechanics had used straps to pull the blade to full lag position without using power (hydraulic & electric).

3. Corrosion Pits;
4. Key ways and
5. Abrupt sectional change.

Inhomogeneity and metallurgical stress risers are given below:

1) Porosity is caused by air bubbles in the mold or in the molten material.
2) Inclusions, segregation in aluminum and/or silicon carbides.
3) Forging errors consist of "laps," "seams,", or "folds."

Examples of microscopic stress risers include grain boundaries and crystalline lattice defects. Stress concentration factor K is used to quantify the severity of a stress riser:

$$K = \frac{\text{maximum local stress}}{\text{nominal stress}}$$

Thus if you calculate the nominal stress at the stress riser, and then multiply the nominal stress by the appropriate stress concentration factor, you arrive at the maximum local stress at the stress riser.

Sources of calculating stress concentration factors are theoretical and experimental. Theoretical analyses are based upon mathematical theory of elasticity and finite element method. The mathematical theory of elasticity may yield "closed from solutions," i.e., formulas for k.

Finite element method may be used to find approximate solutions when you can set up, but not solve, the elasticity problem. Experimental methods can be used when analysis fails.

A well-known example of theoretical solution for stress concentration factor is an elliptical hole in a semi-infinite plate under uniform uniaxial stress. Formula for stress concentration factor K is: $K = 1 + 2\dfrac{(a)}{b}$; the "a" dimension is always across tension. For circular hole: a = b; K = 3, at end of "a" axis.

An application of the elliptical hole formula to analysis of cracks in plates is a heuristic theory only. Actual cracks do not have an elliptical shape. However, the results which we get from assuming their shape to be a very flat ellipse are correct qualitatively, if not quantitatively, are a lot easier to calculate, and will provide us with valuable insights. Two other cases of this formula are:

Crack parallel to applied tensile stress, a=0, a/b=0, K=1

Crack normal to applied tensile stress, b=0, a/b=∞, K=∞.

In real life we always encounter stress concentration either by design or error and mishaps. To reduce the stress concentration factor key concepts include making the discontinuity less abrupt, stop drilling to blunt the crack and blending nicks on structural surfaces. There are techniques for stop drilling the crack comprising the location and size of the drill. Likewise there are, techniques for blending nicks comprising depth and direction of blending the nicks for optimum performance.

There are some useful applications of stress risers. These include, glass cutting, preferred location of failure, tissue paper, reply invoices and check books etc.

There are two commonly used experimental methods for measuring stress raisers, namely, extensometers and strain gages. Each method has its benefits and disadvantages. An experimentalist can apprise the reader of their uses and preferences.

There is a photo elasticity based demo kit which is used in the class to demonstrate the effects of stress concentration on stress distribution in different specimens.

Theory and application of photo elasticity based stress analysis, "Opticon Demo" are given below:

Certain transparent materials have a property called "birefringence". Under stress their optical properties are altered/ so that when polarized light is passed through them it is resolved into components which are aligned with the principal directions; and the velocity of each component is retarded in proportion to the magnitude of this principal stress on its plane.

When the light emerges from the material there is a "phase difference" between the light components, which depends on the difference between the principal stresses and the distance the light has traveled through the birefringent medium. If white light is used, colored bands result, since different light frequencies (colors) are affected differently by the birefringent material.

In our demo we will apply stress in only one direction - compression parallel to the axis of the specimen - and thus the colored lines ("fringes") will be lines of constant stress. Where the lines are close together, the stress is changing rapidly; e.g., where loads are applied, and at stress risers.

Note that our demo shows that where multiple stress risers exist in a location, their stress concentration factors combine by multiplication.

Effects of ductility on stress concentration, at yielding load material relieve stress concentration.

In certain cases ductility may even cause beneficial residual compressive stress.

Brittle materials won't yield, so they break!

Under impact loads, ductile materials may acts as brittle materials.

At low temperatures even ductile materials can become brittle material.

On repeated or fluctuating loads, we'll see that stress risers are important here, too. Stress Concentrations are almost always important in Aircraft Structures.

9

FATIGUE

When a part is subjected to repeated loads, stresses are created within the parts and they continually vary with time. The maximum values of these loads are of significance in the static strength requirements of a structure. However, the cumulative effects of all the various repeated loads are of great importance in the requirements of service life of a structure. It is important to note that stresses less than the static yield point may cause failure if repeated a sufficient number of times. Consequently, a part may fail in a progressive manner due to fatigue.

The behavior of a metal under cyclic stress is fundamental to an understanding of the process of fatigue failure. When a metal sample is subjected to continuous cyclic stress, specific occurrences give evidence of the progressive nature of fatigue failure. At the beginning of the cyclic stressing, fine slip lines develop within the crystalline grains. The mechanism underlying this fine slip is the generation and movement of dislocations within the crystalline structure. The generation of these fine slip lines is evident immediately after the application of the cyclic stressing.

As the cyclic stressing continues, the fine slip lines extend, intersect other slip lines, and become more numerous throughout the crystalline structure. The continuation of this cyclic fine slip creates the first real evidence of deterioration. Small microscopic cracks will begin to form within five or ten percent of the fatigue life of a part. Several possible situations may be used to explain the generation of these microscopic cracks:

(1) The cyclic plastic strain causes a progressive unbonding within the atomic lattice and initiates the beginning of a microscopic crack.

(2) It can be assumed that all ordinary materials may contain submicroscopic cracks since these are not perfect crystals. The generation and movement of dislocations into the vicinity of these submicroscopic cracks will intensify local stresses and cause disbonding or enlargement and growth of the cracks.

(3) The piling up of dislocations in certain critical areas may allow a condensation to form microscopic cracks (i.e., if enough dislocations are pushed together, a vacancy or crack is created).

Which of these possibilities actually exist is probably determined by the nature of the material and the exact type of stressing. In any case, it is important to note that microscopic cracks are formed early in the fatigue life.

After the initial appearance of microscopic cracks, continued application of cyclic stress simply causes a growth in the number and size of microscopic cracks. This subsequent deterioration leads to the joining and rapid growth of a large crack in some critical area. This critical macroscopic crack will spread rapidly and become visible by ordinary means of inspection such as penetrant dye, x-ray, etc. The appearance of the final critical macroscopic crack will occur at approximately sixty to eighty percent of the fatigue life. Up to this point, all plastic strain has been highly localized fine slip and there will be no appreciable distortion or deformation in the vicinity of the final critical fatigue crack. As a result, the failure surface in this vicinity will appear quite brittle, even though the metal would demonstrate a ductile behavior under static loads.

The final visible crack will become apparent in the last twenty to forty percent of the fatigue life. The next phase of failure is the propagation of the critical crack into the cross-section of the part. As the crack spreads into the cross-section, some point is reached where the remaining cross-section cannot withstand the imposed load. At this point, a final sudden rupture occurs on the remaining cross-section. Of course, this final rupture will take place with considerably more distortion than the previous crack formation and will be more typical of a static, stress-rupture type of failure. This characteristic sequence produces the most identifying feature of a failure due to fatigue. Two types of fracture surfaces are evident: the relatively brittle area of crack propagation and the relatively ductile and distorted area of final rupture.

The sequences of events which occur during a fatigue failure allow one to draw some specific conclusions about the nature of fatigue:

(1) A fatigue failure is a progressive failure.

(2) The static strength of a part is essentially unchanged until the final critical crack is formed and propagated.

(3) A final critical crack will be formed and propagated until rupture occurs on the remaining cross-section. A crack will always be formed prior to the final rupture (there is no such thing as a sudden crystallization or instantaneous fatigue failure).

(4) The damage or deterioration that occurs during fatigue stressing is accumulated and is irreversible, except by complete reformation of the material.

(5) Fatigue is due to the accumulation of cyclic plastic strain. In addition, the predominate features of typical service fatigue failures show that there must be a tensile stress present to propagate the crack.

Derivation of the fatigue strength properties of a material is more difficult than obtaining its static strength properties. Many more samples are required and considerably more time and equipment are necessary. The usual means of a fatigue test is to subject a sample to continuous application of a particular stress until failure occurs. Of course, many samples must be used to adequately investigate the available range of stresses for the material under test. The results of a fatigue test are displayed as a graph of cyclic stress (S) versus the number of cycles (N) to produce failure. A typical S-N diagram will most likely display a logarithmic scale for the number of cycles (N) for two reasons: (1)

any load application of less than one cycle would have no significance and (2) since the number of cycles at low stress may be very large, a condensing of the scale is favorable to an accurate portrayal of material properties. A representative S-N diagram is illustrated in Figure 9.1.

The specific properties illustrated in Figure 9.1 are for a typical high strength aluminum alloy. The static strength properties of this material are a tensile ultimate strength of 82,000 psi and offset yield strength of 70,000 psi. The diagram of Figure 9.1 illustrates the only real correlation between static and fatigue properties at one cycle of load. One cycle loading is certainly not fatiguing and the material can withstand the full 82,000 psi for this load application. The important distinction for fatigue is noticed at cycle loading of 100,000 or 10^5. At this point, the specimens of material subjected to a cyclic stress of 57,000 psi would fail beyond 100,000 cycles of this stress. Of course, reduced cyclic stress requires considerably greater number of cycles to cause fatigue failure (e.g., stress of 35,000 psi could be applied 500,000,000 cycles prior to failure).

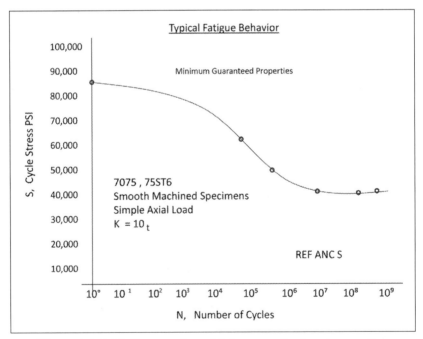

Figure 9.1 S-N diagram for a high-strength aluminum alloy

The important fact to be noted is that the material may be failed in fatigue at stresses well below the allowable static stresses if sufficient cycles of stress are applied. A more general correlation between static strength and fatigue strength can be made by use of Figure 9.2. The upper illustration is a comparison of the typical static stress-strain (σ (f) vs. ε (e)) diagram and the fatigue S-N diagram. The only exact point of comparison occurs at the one cycle point of ultimate strength σ_u. Since the yield point and proportional limit are defined by an offset method, it is implied that a certain small plastic strain exists even at very low stresses. This fact is given support by the corresponding S-N diagram in that the allowable cyclic stress continually decreases with increasing cycles of stress. While aluminum alloy is mentioned, this does not exclude other materials since this behavior is typical of most metals in general.

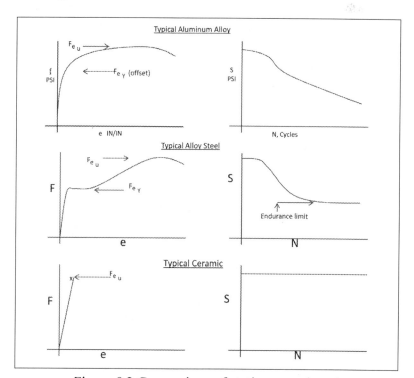

Figure 9.2 Comparison of static strength and fatigue strength graphs for three materials

A special situation is depicted in the second illustration of Figure 9.2 which depicts a typical correlation of static and fatigue properties of a representative alloy steel. The typical stress-strain diagram of steel is characterized by a distinct yield point as well as a distinct proportional limit. The elastic behavior of the steel at low stress is distinct and typified by a straight line up to the proportional limit (no offset). Thus, typical steel will demonstrate static and dynamic elastic behavior below the proportional limit. As a result, the S-N diagram for the steel becomes asymptotic to some stress at the upper level of the apparent elastic range. This particular stress is referred to as the *endurance limit* and if never exceeded in operation, fatigue failure will not occur. Steel is the only metal which demonstrates this particular behavior, implying certain superiority for steel in structures which require very long fatigue life.

Another special case is illustrated in the lower illustration of Figure 9.2 by comparison of static and fatigue properties of a typical ceramic material. Most ceramics at low temperature demonstrate nearly perfect brittle behavior (i.e., the stress-strain diagram is a straight line to the point of fracture without plastic deformation at any point). Fatigue tests of such materials result in S-N diagrams with horizontal lines corresponding to the fracture stress under static conditions.

The illustrations in Figure 9.2 simply amplify the fact that fatigue failure is due to the accumulation of cyclic plastic strain.

All of the previous fatigue properties depicted have been applicable for a specific nature of stressing: the cyclic application of a simple axial tensile stress. It should be obvious that any change in the manner of loading and the resultant change in the actual type of cyclic stressing can create considerable change in fatigue behavior. The particular effects of a specific type of loading must be noted for each manner of fatigue loading (e.g., rotating beam, axial load, reversed cantilever bending, etc.). In addition, a particular sense must be given to the exact nature of the fatigue stressing. Figure 9.3 illustrates the manner of this stress notation.

The first illustration of Figure 9.3 depicts a cyclic variation of stress with time. The cyclic stress amplitude (σ_a) is one half the entire range of stress variation and is the most important single item describing a situation of fatigue stressing. The average or mean stress (σ_m) is the particular bias of stress upon which the cyclic stress is imposed.

The significance of the mean stress is illustrated by the second illustration of Figure 9.3 which illustrates the fatigue behavior of a moderate strength aluminum alloy. As in previous fatigue diagrams, cyclic stress (σ_a) is plotted versus the number of cycles required to produce failure (N). In addition, the effect of mean stress is illustrated by lines of fatigue properties for s_m from -10,000 to +30,000 psi (Note: + denotes tension while - denotes compression). Figure 9.3 illustrates a feature which is typical of all metals: an increasing bias of tensile stress is seriously damaging with respect to fatigue life

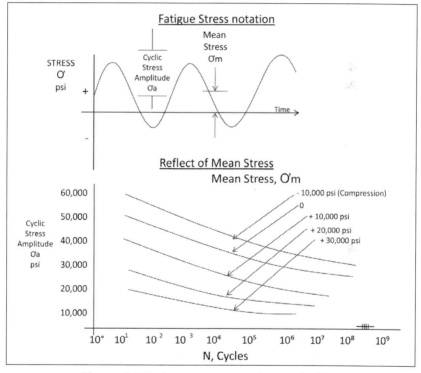

Figure 9.3 Fatigue stress notation and effect
of mean stress on an S-N plot

In this respect, tension structures will be most prone to fatigue while compression structures will be relatively free from fatigue. An alternate method of describing the effect of alternating and mean stress is to plot allowable cyclic stress versus mean stress for a fixed number of cycles.

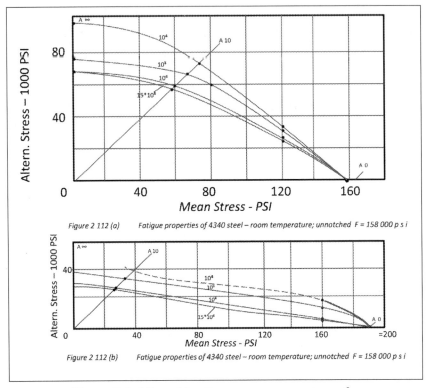

Figure 9.4 Allowable stresses vs. mean stress for
unnotched and notched 4340 steel specimens.

Figure 9.4 illustrates this method by presentation of some typical fatigue data. There is an inevitable variability of material strength properties. This variation of properties may be considerable in some instances such as castings, welded joints, cross grain stressing, etc. and necessitate special consideration during design and manufacture. The primary structural elements of aircraft and missiles require a high level of quality control to insure that the lower limits of material properties are well established to allow precise minimum weight design.

The effect of the normal variation of material properties is illustrated in Figure 9.5. Suppose that a group of ten specimens are subjected to fatigue stressing at a particular level of cyclic stress: S1, S2, or S3. The normal variation in material properties will mean that there will be a variation in the number of cycles at which each of the ten specimens will fail due to fatigue. Under certain conditions, it may be that the last specimen will withstand ten to one hundred times the cycles necessary to fail the first specimen.

The group of specimens will be distributed or scattered about a median in some fashion, which would be best analyzed by use of probability theory. It would then be logical to define the fatigue characteristics of a material by lines of failure probability on the S-N diagram. This is the usual means of accounting for the scatter of data points resulting from fatigue tests. Such a presentation is illustrated by the second chart of Figure 9.5. If a group of parts were subject to a particular level of fatigue stress, the probability of fatigue failure would be defined at the appropriate cycles corresponding to various values of probability. If sufficient cycles were applied to a group of specimens to create a probability of 0.5, a liberal interpretation would relate that 50% of the specimens would be failed due to fatigue at this point.

At this point, it is appropriate to relate some of the fundamental ideas utilized in design analysis to insure a fatigue-free service life. During service operation, a structure may be subjected to a variety of loads during which the stress amplitude, mean stress, and cyclic application may vary. As recalled from the description of the fatigue process, the effect of continued cyclic stressing is the accumulation of cyclic plastic strain. From the beginning of cyclic stressing with the generation of fine slip and microscopic cracks to the final phases of a critical crack growth and final rupture, the entire sequence represents the progressive accumulation of fatigue damage.

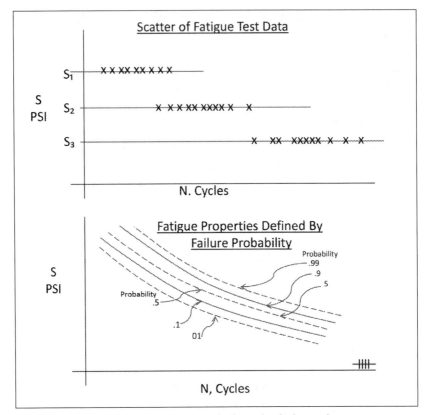

Figure 9.5 Typical variations in fatigue data

The concept of cumulative damage is an important one in analyzing a structure for the effects of the spectrum of loads encountered during service operation. A simplified version of the cumulative damage concept is presented by the illustrations of Figure 9.6. Suppose that a particular part is subjected to n_1 cycles of cyclic stress S_1, n_2 cycles of S_2, n_3 cycles of S_3, etc. The S-N diagram (of appropriate probability) for the material depicts an allowable cycle N_1 when S_1 is the only cyclic stress imposed on the part. Similarly, the cyclic stress S_2 has allowable N_2 cycles, S_3 has N_3 cycles, etc.

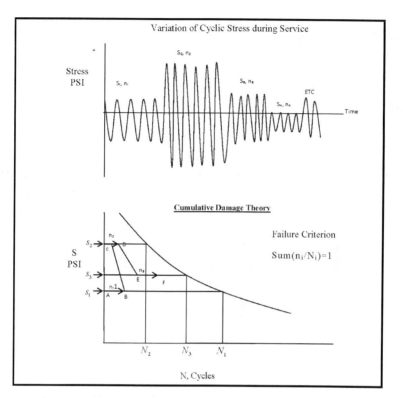

Figure 9.6 Example of cumulative damage from fatigue

During the spectrum of loading, the application of n_1 cycles of stress at S_1 will utilize a fraction of fatigue life which is represented by the proportion of n_1 (applied) to N_1 (allowable). The subsequent application of n_2 cycles of stress at S_2 will utilize an additional increment of fatigue life represented by the proportion of n_2 and N_2. Thus, the application of n_1 cycles at S_1 would accumulate damage from points A to B on Figure 9.6. Subsequent application of n_2 cycles at S_2 would accumulate damage from points C to D, and if n_3 cycles at S_3 are added, damage would accumulate from points E to F.

The criterion for fatigue failure under variable fatigue stresses is represented by a critical accumulation of damage. When the sum of the fractions of damage is unity, a critical damage has accumulated and fatigue failure occurs. This concept of damage is represented by the Miner hypothesis of cumulative damage:

$$\sum \frac{n}{N} = 1 \quad or \quad \frac{n_1}{N_1} + \frac{n_2}{N_2} + \frac{n_3}{N_3} \cdots = 1$$

The use of this fundamental concept is required to insure fatigue-free operation of a structure within the anticipated service life at a particular spectrum of loading.

There are many particular aspects to be accounted for during fatigue design. The load spectrum must be identified, stress concentrations must be held to a minimum, material properties must be derived, and the operational service life of the parts must be anticipated. In addition, some very fundamental aspects must be given due regard. During design, the structure may be shaped such that the accumulation of fatigue damage during the anticipated service life does not cause fatigue failure. At a minimum, the probability of failure can be set to such a low level as to insure satisfactory service performance of the structure. An alternate method would be to require specific fatigue factors of safety based on an equivalent fatigue stress which represents the effect of the load spectrum and desired service life.

Regardless of the specific technique of analysis employed in design, there will be great difficulty in estimating the fatigue life of a single component. The normal scatter encountered in fatigue tests would allow the prediction of fatigue life of specific components only on the basis of failure probability. Thus, if one hundred copies of a particular part are manufactured and placed in service, the specific fatigue life of one of these units cannot be defined with suitable accuracy. However, the probability of failure within these one hundred units can be assessed as a function of the accumulated service. This fact alone is sufficient to justify meticulous inspection and overhaul of systems on a regular and continuing basis, especially if component use is extended well into the original anticipated service life.

One important fact with regard to fatigue is that there must be a full appreciation of the manner in which fatigue damage is accumulated. Fatigue is due to the accumulation of cyclic plastic strain and is related to the number of cycles of an applied fatigue stress. The actual time of accumulation is of very minor significance. The frequency of the cyclic stressing is not an important factor until the frequency is so high as to limit the plastic straining during cyclic stressing. Ordinarily, the very high frequencies required to cause a significant change in fatigue behavior are encountered only in rare instances. Thus, a critical number of fatigue cycles may be applied in a few hours as well as a few hundred or thousand hours, with the effect on the material being essentially the same in any case. As a result, the accumulation of fatigue damage must be appreciated as the accumulation of cycles of stressing rather than time since manufacture. The cycles of stressing and time in service are related only by the particular load spectrum.

In addition, the overhaul and inspection of parts which have been in service will guarantee only that the parts are fully capable of original static strength. Suppose a service item is thoroughly overhauled and inspected and found to be free of cracks. This situation does not guarantee that prior accumulation of fatigue damage would not allow crack formation when returned to service. Since the final critical crack originates and propagates during the last twenty to forty percent of life, a service inspection could guarantee no more than twenty percent of the service life as failure free operation. Also, the damaging effect of short periods at high cyclic stress would tend to cause more rapid crack propagation. For this reason, it would be wise to assume that a service inspection be accommodated at periods which are less than twenty percent of the anticipated service life.

Fatigue is caused by the accumulation of cyclic plastic strain. The rate of fatigue damage accumulation is increased significantly by the presence of tensile stresses (refer to Figure 9.3). Any attempt to improve the fatigue resistance of a structure must involve reducing the cyclic plastic strain and reducing the magnitude of the tensile stresses which may be present. The effect of alloying and heat treatment is primarily to harden the material and increase the resistance to plastic flow. The creation of solid solutions, particle precipitation, and dispersion of second phases can stabilize or immobilize dislocations and increase the resistance to plastic strain. Strain hardening, when applicable, will extend the elastic behavior to higher stresses. In addition, cold working may produce a desirable grain direction and give increased fatigue resistance along the grain direction (but reduced fatigue resistance and greater scatter when stressed across the grain).

Stress concentrations of all sorts are highly damaging in fatigue since the action of the discontinuity is to cause a serious local increase in cyclic stress and strain. Certain stress concentrations are inevitable in manufactured structures. Rivet and bolt holes and changes of cross-section are unavoidable at attachments and joints, and contribute an undesirable effect on fatigue life. Of course, all effort must be made to insure that the resulting stress concentration is at a minimum by providing smoothly finished holes and changes of section with generous fillet radii. The surface finish of a part contributes to the fatigue life by reducing the surface stress concentrations of tool marks, forming lines, pits, flaws, and fissures. A smooth, polished surface will contribute greatly to fatigue resistance.

Experience has proven that the majority of fatigue failures encountered in service originate at a distinguishable stress concentration. In some instances, the stress concentration was the result of inadequate detail design or poor quality control during manufacture. Other cases have shown the introduction of additional stress concentration resulting from foreign object impact, poor handling and maintenance practices, or corrosion. Since the surface of a part (rather than the interior) is more prone to the action of stress concentrations, favorable residual stresses of compression can be introduced to increase fatigue resistance.

An aircraft propeller is a typical structure subjected to a variety of fatigue loads and an environment which is likely to cause considerable surface stress concentrations (e.g., nicks, pits, and scratches) due to impact with foreign objects. In most instances, the critical areas of the propeller blade will be shot-peened to provide a favorable residual stress of compression at the outer surface. The effect of shot-peening is illustrated in Figure 9.7 in the simplified case of a part subjected to simple axial loading. An original part subjected to a simple axial tension load develops a uniform tensile stress across the cross-section as seen in (A). When the part is shot-peened, extensive compressive plastic deformation occurs on the outer surface and a residual compressive stress is developed in the outer surface, as depicted in (B). Of course, a small residual tensile stress is developed within the interior. As the shot-peened part is subject to an axial load in (C), the resultant stress distribution is the sum of the uniform axial stress in (A) and the residual stress distribution in (B).

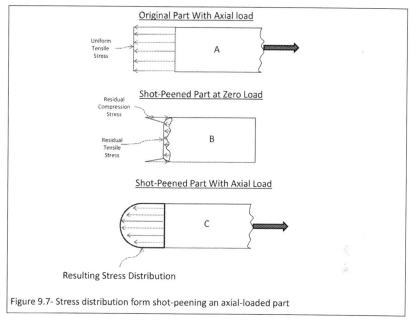

Figure 9.7- Stress distribution form shot-peening an axial-loaded part

Figure 9.7 Stress distribution from shot-penning an axial-loaded part

The principal effect is that the net tensile stress at the outer surface is reduced and the resistance to fatigue is increased. The net tensile stress on the interior is increased slightly, but this is of no real concern since the exterior surface is more vulnerable to the action of fatigue and stress concentration.

The highly beneficial effect of residual compressive stress is utilized in many ways. Bolts with threads and head-to-shank fillets which are cold-rolled have high residual compressive stresses in areas of high-stress concentration. The fatigue resistance of such bolts is superior to other bolts which have hot-rolled or machine cut sections. In the same sense, lap joints which are joined by preloaded, tapered pins have a residual compressive stress developed around the pin hole. The preloaded, tapered pin bears against the sides of the hole, inducing the favorable residual compressive stress in the area of critical stress concentration. The improvement in fatigue resistance is considerable, but the precision and cost of such a joint will ordinarily limit the application to particularly sensitive sections.

It is an established fact that ductile metals do not experience any significant loss of static strength from the action of stress concentrations. The ductile metal will develop large plastic strain in the vicinity of stress concentrations, and this will tend to redistribute the stresses to produce a reasonably uniform stress distribution in the plastic range. When the applied load is released, the plastic distortion previously accommodated at the stress concentration will produce a residual stress. If the applied load produces an initial tensile stress at the discontinuity, release of the load will incur a favorable residual compressive stress at the stress concentration. Thus, proof testing to limit load of a ductile structure prior to service could improve the fatigue resistance. Of course, continuous application of high proof stresses during service inspection and testing could be quite damaging in respect to fatigue life.

When fatigue failures occur during service operation, a thorough inspection of the failed part is necessary to prevent repetition of failure. Accurate maintenance and periodic inspection is necessary to allow location of incipient failures prior to final rupture. Maintenance must insure that no undue stress is developed due to improper rigging, lack of proper preload, or shock and vibration from excessive clearances. Structures subject to fatigue-type loads must not be subject to poor handling, foreign object impact, poor shop practices, or any other environment that might develop excessive stress or surface discontinuities. In the case of a failure, the part must be examined to define any deficiency of the basic material hardness. Perhaps the failed part has been subject to higher than ordinary temperatures and a loss of heat treatment or favorable residual stress has occurred.

The rate of fatigue damage accumulation is most rapid when a part is subjected to high cyclic stresses of tension. For this reason, short periods of high overstress can cause a serious accumulation of fatigue damage.

9.1 Commonly Used Terms

Definition: Fatigue is the progressive localized permanent structural damage that occurs in a material subjected to repeat or fluctuating, strains at stresses having a maximum value less than the tensile strength of the material. After a sufficient number of cycles fatigue may result in cracks or fracture in the part.

It has been estimated that 80% of *operating* failures of machine parts (rotating components) are due to fatigue loading. First cracks don't appear until 40-60% of fatigue life has been expended. It's a major problem!

Requirements for fatigue are defined by the "Fatigue Triangle", with following arms:

1) Cyclic Stress **initiates the crack,**
2) Tensile Stress **propagates the crack, and**
3) Plastic stress **arrests the crack**.

Can we eliminate any of the above conditions of fatigue failure in structures? We cannot eliminate plastic deformation and cyclic loads. Tensile load can be avoided by shot peening. Shot peening introduces residual compressive stress on surface of part, where stress risers occur.

Critical Crack Length: As crack grows, stress concentration factor grows (a/b of the assumed elliptical crack grows) finally reaches maximum length for crack to be arrested. Next time load comes on, crack runs to boundary. Progressive damage takes time, first crack must initiate, and then propagate and only then can you have fracture.

Examples of Aircraft mishaps:

Argentine HS-748 mishap from Aviation Week – 34 fatal

Royal Aircraft Establishment, Farnborough Laboratories showed critical crack length of 36" took 10,000 flight hours to grow.

Investigations found that another 30" long with 0.007" width crack in another area.

Aloha Airlines mishap, http://www.aloha.net/~icarus/.

F-15 mishaps and grounding of fleet for few days was reported in press. http://www.latimes.com/nation/la-na-f156nov06-story.html

A Google search on relevant topics will provide many other cases.

Fatigue life determination using fatigue testing: Repetitive cycle testing was the first technique developed by British Railways. First fatigue testing was based upon rotating bending tests on rail road axles of coal cars in British Railways in 19th century. Now it is well known technique for fatigue life determination of aircraft materials, components and structures.

Constant amplitude loading: Under simulated service testing Miners rule makes assumptions that there exists no dependence on number of cycles per hour or on duration of load. Thus, we can compress multiple lifetimes into a year or two of testing. Applications with more complicated loadings need research. But magnitude and order of loading are important. Full scale fatigue tests to 2x lifetime thus "For satisfactory correlation with service behavior, full-size specimens must be tested under conditions as close as possible to those existing in service." Per Metals Handbook Block loading schemes are used to group similar loads for convenience in testing. Thus one doesn't have to reset controls as often. With use of digital computers, Testing of "Ground-Air-Ground" cycles, does not affect the result because of the order of loading. The results of test specimens in terms of stress vs nunmber of cycles to failure are called "S-N Diagram". S-N curve is also known as Wohler curve or diagram. Figure 9.3 shows stress level vs fatigue life for a representative material; "n" = fatigue life = cycles of failure in log scale. It is shown that s-n curve data follows Weibull statistics.

Endurance limit of a material is stress level below which the fatigue life is infinite. N=infinity.

Fatigue strength of a material is stress level at which half of the specimens have failed (and half are still going) after a given number of cycles. Aluminum doesn't have an endurance limit.

Typical factor of safety on life is 2 for data and 4 for model.

Miner's Law: Diagrams such as the S-n diagram present fatigue life predictions for constant-magnitude cycles (back and forth, from max to min) from new part to fatigue failure. This does not consider order of loading. Actual service involves widely varying cycles in apparently random order. Miner's Law is a way to combine the effects of different loadings to calculate the total damage done. Mathematically we can state that:

$$\sum \frac{n_i}{N_i} = \text{FRACTION OF LIFE USED. } =1 \text{ at failure,}$$

Where n= actual applied cycles and N=life at corresponding load.

Computation procedure: For each load condition, find the fraction of life used by dividing the ni, the number of cycles of load condition, "i" actually experienced by Ni, the fatigue life under load condition "i" from new to fatigue failure. Add up all the fractions formed and find the total fraction of life used by the combination of loads experienced.

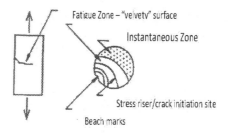

Figure 9.8: damage description in fatigued specimen

Fatigue Stages and Zones: The following example Figure 9.8 gives different stages and zones in fatigue tested specimens:

Stage/Zone 1: **Crack starts** at a location of maximum stress and minimum strength which is mostly at stress riser. In ductile material crack orientation is usually at 45° to the normal stress, on a "slip plane," a plane of maximum shear stress.

The size of initial crack is never more than two to five grains – i.e., microscopic. This small damage is easy to happen and this zone cannot be seen with naked eye. It requires 200X magnification.

Stage/Zone 2: **Crack propagates** as parallel plateaus normal to the maximum normal (tensile) Stress. The surface texture is velvety smooth but not shiny like "machined" shear lip. These, appearance and orientation, are the clues of ductile material fatigue damage and 80% correct.

Other causes for crack propagation include:

Chemically active surface change frequency or magnitude of cyclic load,

High cycle fatigue – one surface polishes the other!

Fatigue striations (submicroscopic damage requiring SEM to inspect) and

Plastic strain arresting the crack shows stretch marks of finely spaced lines.

Positive Evidence of Fatigue: Since striations occur one per cycle, which advances the crack, counting striations, allow determination of the number of cycles to failure. With some non-trivial work with the loading spectra for the aircraft, you can convert fatigue cycles in to flight hours!

Ratchet Marks are macroscopic in nature and appear as ledges between the parallel plateaus. The causes of ratchet marks are irregular loads and/or material variability.

Stage/Zone 3: **Instantaneous failure occurs** at critical crack length. For ductile material one finds a shear lip with gross plastic deformation, while brittle material one finds small shear lip and no gross plastic deformation and granular surface.

"Crystallization."

Fully reversed loading cause shiny surface whereas re-crystallization phase occurs at elevated temperature.

High cycle vs. low cycle fatigue is important aspects to understand the response characteristics of materials and their appropriate application. This results in to high overstress vs. low overstress fatigue?

Recognition of Fatigue Failure:

Every fatigue failure has at least two distinct *"unrelated"* zones: at least one Fatigue zone, and the Instantaneous zone. The distinction may not be obvious in brittle materials.

The fatigue zone in brittle materials always resembles a brittle fracture: the fracture surface is oriented normal to tension, and there is no GPD.

The fatigue zone is oriented normal to the maximum normal stress (tension), even in a ductile material.

The fatigue surface usually has a velvety, satiny, light-reflecting surface, distance from the normally rough or granulated surface of the instantaneous zone.

Beach Marks indicate the progression of the fatigue crack across the fatigue zone.

Concave beach marks show the point of origin of the fatigue process.

Ratchet Marks frequently result from non-homogenous material and/or high load levels (low cycle fatigue).

Normally the instantaneous zone has all the appearance of the static type of failure, which results from the type of loading which caused the failure. Thus the type of loading involved may usually be determined from the appearance of the instantaneous zone.

The degree of ductility of the part is indicated by the percentage of the instantaneous zone, which is occupied by the shear lip.

RULE OF THUMB: If you can't see the shear lip with the naked eye, the material is either very brittle or was failed by a high impact load. The failure analyst is able to determine which, of course!

High cycle/low cycle (low overstress/high overstress) fatigue is indicated by the relative amount of the total fracture surface which is occupied by the fatigue zone (in a uniformly heat treated material).

10

COMPOSITES

Composites are receiving increasing attention by product designers because of their versatility of properties, tailor making nature and manufacturing ease.

Composite Material is a combination of dissimilar materials in which each constituent remains identifiable, but in which the mechanical composite are different from the properties of any one of the constituents of which it is comprised.

Advanced Composite is a material consisting of high-strength, high modulus, low-density fibers embedded within a compatible, essentially homogenous matrix.

Filamentary Composite is a composite material in which reinforcing fibers are in the form of filaments: long, continuous strands, the majority of which do not end within the part (the ends are at the edges of the part).

Laminated Composite: a material made by bonding together two or more layers ("laminate") of a material or materials. In advanced laminated composites the orientation of successive layers is changed to produce the desired material properties in the laminate.

Because of the high strength and low weight characteristics of composite materials, they are most attractive for aircraft and air space structures.

There is a potential of the following three problems, which may occur when using composites in aircraft structures:

a. Low-energy impact: Low energy impact of composites may cause invisible damage, leading to matrix, fiber and delamination failure.

b. Fire: Polymer composites under fire can be very hazardous. In addition to the aggravated fire there is a severe danger of flying fibers from the composite structure.

c. Some Corrosion: The insertion of water may cause deterioration of matrix, interface and/or fiber.

Some additional special problems which may confront the investigator when working on a mishap involving composite structures are:

1. Splinters
2. Smoke
3. Released fibers
4. Smoldering combustion
5. Failure analysis

Additional benefits of composites are given below:

1. Higher specific strength, which lead to lighter structures
2. A lower density, which saves weight
3. The ability to tailor laminate patterns to achieve desired stiffness and strength
4. The ability to manufacture stable components having zero coefficient of expansion
5. The opportunity to improve an original design during the process of changing the material
6. The elimination of corrosion
7. Higher resistance to fatigue damage
8. Lower manufacturing costs by reduced machining
9. Absorption of radar microwaves (stealth capability)
10. The significant "high-tech" sales appeal

10.1 Composites vs Metals:

10.1.1   Properties are not uniform in all directions
10.1.2   Strength and stiffness can be tailored to meet loads
10.1.3   Greater variety of mechanical properties

10.1.4   Poor through-the-thickness strength

10.1.5   Greater sensitivity to environmental heat and moisture

10.1.6   Greater resistance to fatigue damage

10.1.7   Propagation of damage through delamination rather than through-thickness cracks

10.2 Advantages of Composites over metals:

10.2.1   High resistance to fatigue

10.2.2   Reduced machining

10.2.3   Tapered sections and compound contours easily accomplished

10.2.4   Tailoring of fibers' orientation to meet strength/stiffness directions requirements

10.2.5   Reduced number of assemblies when co-curing or co-consolidation is used

10.2.6   Light weight

10.2.7   Absorption of radar microwaves (stealth capability)

10.2.8   Thermal expansion close to zero, reducing thermal stresses in space applications

10.3 Disadvantages of Composites vs Metals:

10.3.1   Material is expensive

10.3.2   Lack of established design allowables

10.3.3   Corrosion problems can result from improper coupling with metals, especially when carbon or graphite is used (sealing is essential)

10.3.4   Degradation of structural properties under temperature extremes and wet conditions

10.3.5   Complex inspection methods (reliable detection of defects in bonds is difficult)

10.4 Notations for ply stacking directions and laminate sequencing:

10.4.1   Each ply is shown by a number representing the direction of fibers, in degree, with respect to a reference axis. The 0° fibers of either tape or fabric are normally aligned with the largest axial load

10.4.2   The plies are listed in sequence, set off by brackets, starting from the side indicated by the code arrow. For example [0/90/±45] represents a laminate with plies stacked with 0 degree ply at the top followed by a 90 degree ply, then a 45 degree ply and then a -45 degree ply.

10.4.3   Adjacent plies with different angles of orientation are separated by a slash. For example [0/90] represents a laminate with top 0 degree ply and bottom 90 degree ply.

10.4.4   Callouts of fabric plies are differentiated from tape plies by parentheses

For example [(±45)/(0,90)] represents a fabric with 45 degree and -45 plies; and fabric with 0 degree and 90 degree plies. An example of hybrid (fabric and tape) laminate is [0/(±45)/90] which represents a 0 degree tape; 45 degree fabric and -45 fabric; and 90 tape plies.

For tape only, when ± is used, two adjacent plies are indicated, with the top symbol being the first of the two plies. For example [±45] represents 45 degree and – 45 degree plies stacked together.

10.4.5   Adjacent plies of the same angle of orientation are shown by a numerical subscript.

For example [0/90$_3$/0] represents a laminate with 0 degree top ply, followed by three 90 degree plies and bottom 0 degree ply.

A symmetric laminate is shown by using the subscript "s" following the stacking sequence.

For example an eight ply symmetric laminate [0/90/±45]$_s$ represents

[0/90/+45/-45/-45/+45/90/0] stacking sequence.

10.4.6   Symmetric laminates with an even number of plies are listed in sequence, starting at one face and stopping at the mid-point (the plane of symmetry). A subscript "s" following the bracket indicates that   only one-half of the code is shown, with the other half being a mirror-   image. We can use the example shown above [0/90/±45]$_s$.

10.4.7   Symmetric laminates with an odd number of plies have the center one over-lined to indicate this condition. Starting with this ply, the rest of the code would be a mirror-image to that part shown. $[0 / (\pm45) / 90']_s$ can be used as an example with odd number of plies. A sample of representative laminates is given below.

10.5 Macro-mechanics:

A science of computation of average apparent properties of composites (gross behavior of composites e.g. laminate analysis) from ply properties is termed as Macro-mechanics. Using the ply direction dependent properties with appropriate transformation relations along laminate ply angles and weight/thickness averaging over laminate ply orientations, laminate properties are calculated. Ply mechanical properties needed can be obtained from predictions using micromechanics (please refer Figure 10.1) models. Sometimes, in case of lack of information about constituents and their properties, experimental values of ply properties are used in computing laminate properties. When predicting failure, the statistical nature of the fracture processes can be difficult to model.

<u>Example:</u>   <u>***Laminate Stacking Sequence***</u>

Unidirectional: $[0/0/0/0/0/0] = [0_6]$
Cross-ply symmetric: $[0/90/90/0] = [0/90]_s$
Angle-ply symmetric: $[45/-45/-45/45] = [45/-45]_s = [\pm45]_s$
Angle-ply asymmetric: $[30/-30/30/-30/30/-30/30/-30] = [30/-30]_4 = [\pm30]_4$
Quasi isotropic: $[0/+45/-45/90]$
Symmetric Quasi-isotropic: $[0/+45/-45/90/90/-45/45/0] = [0/\pm45/9C]_s$
Multidirectional: $[0/+45/30/-30/45]$
Hybrid: $[0^k/0^k/45^c/-45^c/90^G/-45^c/45^c/0^k/0^k]_T = [0_2^k/\pm45^c/90^G]_s$

<u>Define</u>: S=Symmetric, K=Kevlar, C=Carbon, G=Glass, T=Total

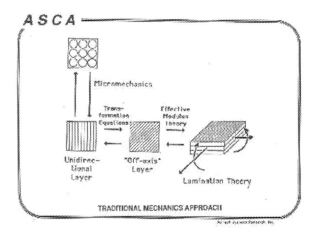

Figure 10.1: micro and macro modeling for composites

10.6 Micromechanics:

Micromechanics deals with the mechanics of material at its constituent level. Using the mechanical properties of constituent, fiber, matrix and particles, one can determine the effective properties of the ply. Using effective properties, one can find the response characteristics of the laminate. Further, micromechanics is useful for:

10.6.1 assessing the potential of new fiber/resin combinations

10.6.2 predicting composites behavior

10.6.3 testing the relevance of particular parameters in design

10.6.4 choosing a material for a defined purpose

10.6.5 assisting in the search for innovative solutions to materials' problems

(It must be noted that macro-mechanics is of little value for examining the effects of fiber or matrix properties.)

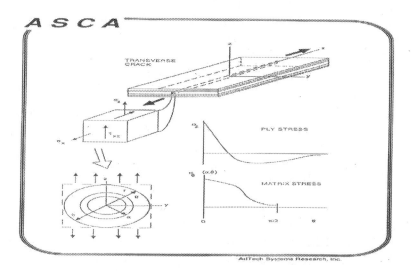

Figure 10.2: Stress calculations for prediction
of matrix and fiber interface failure.

The following books are good references to help one to understand the mechanics of composite materials:

Steven Tsai and H. T. Hahn, "Introduction to Composite Materials"

Stephen Tsai, "Theory of Composite Design"

G.H. Holister and G. Thomas, "Fiber Reinforced Materials"

Sanjay K. Mazumdar, "Composite Manufacturing- Materials, Product, and Process Engineering"

R.M. Christensen, "Mechanics of Composite Materials"

N.J Pagano, " Interlaminar Response of Composite Materials", Elsevier, 1989.

There are many more books in addition to these that deal with composites, written by both the above and other experts in this field of study.

10.7 3D Printing Manufacturing of Composites

3D printing (3DP) is being used extensively to prototype and manufacture various products, however, utilization of 3D printers for aerospace composite applications is quite limited due to a lack of capabilities. A brief overview addressing the relevancy of this technology of relevant technology for understanding the response of composite laminates and structures using 3D printer technology follows. The currently available models (i.e. ASCA, FESAP, ALTRACK and BRAIDS) are considered to evaluate the response of composites in conjunction with the 3D printer technology capabilities. This will conclude with a recommended set of laminate configurations to be modeled, 3D manufactured, and tested for ascertaining the accuracy of predicted response characteristics.

3D printing manufacturing is revolutionizing the manufacturing industry. Digital computing, prototyping and small scale manufacturing is dramatically increasing in the automobile industry, medical industry and food industry. Every day, new applications are being developed for applying this technology. Composite material manufacturing is a potential application that could provide cheaper, stronger, and more durable aerospace components. Current composite manufacturing techniques include ply lay-up using computer aided design (CAD) processes. Those applications are examined below.

Some recent examples of developments in innovative 3D manufacturing that apply to this research include the following (further details on these studies are available through use of Google search on 3D printing of composites):

10.7.1 3D Printed Cement Composites:

Gregory John Gibbons, Reuben Williams, Phil Purnell, and Elham Farahi, "3D Printing of Cement Composites", Advances in Applied Ceramics, 2010, Vol.109 (No.5). pp. 287-290. The aims of this study was to investigate the feasibility of generating 3D structures directly in rapid-hardening Portland cement (RHPC) using 3D printing (3DP) technology. 3DP is an Additive Layer Manufacturing (ALM) process that generates parts directly from CAD in a layer-wise manner. 3D structures were successfully printed using a polyvinyl alcohol: RHPC ratio of 3:97 w/w, with print resolutions of greater than 1mm. The test components demonstrated the manufacture of features, including off-axis holes, overhangs / undercuts etc. that would not be possible to manufacture using simple mold tools. http://wrap.warwick.ac.uk/4465/

10.7.2 3D Printing of Electrically Conductive Materials:

3D printing technology can produce complex objects directly from computer aided digital designs. The technology has traditionally been used by large companies to produce fit and form concept prototypes (*'rapid prototyping'*) before production. In recent years however there has been a move to adopt the technology as a full-scale manufacturing solution. The advent of low-cost, desktop 3D printers such as the RepRap and Fab@Home has meant a wider user base now has access to desktop manufacturing platforms, thus enabling them to produce highly customized products for personal use and sale. This uptake in usage has been coupled with a demand for printing technology and materials able to print functional elements such as electronic sensors.

10.7.3 Canadian Group Claims 3D Printing of Continuous Fiber Composites is Within Reach:

Nathan Armstrong is leading a partnership developing 3D printing technology for the production of continuous fiber composite parts, fed by high-speed, high-efficiency design optimization software. A technology partnership headed up by composites specialist Nathan Armstrong, formerly founder and president of Motive Industries (maker of the Kestrel bio composite car) who has been involved with 3D printing since 1995 and ran a 3D printing company in California, is working to develop a 3D printing technology that will produce composite parts with continuous fiber reinforcement. Further, the technology, says Armstrong, would allow composites design engineers to bypass the traditional design iteration process and thereby bring parts to fruition faster.

One of the limitations of 3D printing to date, as far as composites are concerned, has been that additive manufacturing systems – by virtue of their layered build-up production process — could only produce composite parts reinforced with discontinuous fibers. This limited the application potential of the technology and locked 3D printing out of high-performance markets like aerospace, auto racing, robotics and medical applications. According to Armstrong, 3D printing manufacturing technology to produce composites will be ready by the end of the year 2015.

10.7.4 Mark One 3D Printing of Continuous Fiber Composites:

The Mark One 3D Printer prints carbon fiber, fiberglass, kevlar and nylon. This allows one to take advantage of the non-scratch surface of nylon and the strength of carbon fiber in the same part. Carbon fiber has a very high strength to weight ratio and very high thermal conductivity. Thus it is perfect for applications requiring the greatest possible stiffness and strength. Fiberglass is the most cost-effective material. It is as strong as carbon fiber, but only 40% as stiff and two times the weight. However, it is quite suitable for everyday applications where one needs strong parts. Kevlar has the best abrasion resistance and is our most flexible material for parts that are durable and resistant to impact loading.

This research will focus on fused filament fabrication (FFF) utilizing the Mark One Printer manufactured by Mark Fordged Inc. The emphasis is on understanding the micro and macro mechanics characteristics of composite materials in conjunction with manufacturing of laminates using 3DP technology. We will select the most challenging aspect of composite materials: delamination caused by mismatch of constituent properties. Once we understand this mode of failure in composites, we can develop laminates and manufacturing processes to avoid delamination.

10.8 AS4 Fiber and H3501-6 Matrix Laminates:

Next, we will make a set of laminates prone to delamination using 3DP. A Representative Volume Element (RVE) with fiber and matrix will be considered. Using a sample composite material consisting of AS4 fiber and H3501-6 matrix, effective ply properties of AS4/3501-6 composite are calculated. The model called, "Automated System of Composite Analysis"[1], developed by the Air Force Research Laboratory (AFRL) has been used to analyze aircraft grade composite laminates. Four cases of different Volume Fraction (V_f) of fiber were considered for predicting the ply properties using the RVE model. These values of V_f are .55, .33, .22, .16.

Using the sequence of composite laminate layup as shown in earlier Figure 10.1, we will start with the selection of fiber and matrix materials. Fibers are embedded in matrix as shown at the top element of Figure 10.1. Circles represent the tows of the fibers surrounded by matrix. Based upon the properties of the fiber and matrix materials given in Table 1, ply properties are computed. For the case of Vf=.55, computed ply properties are given in Table 2. Figure 10.2 shows the sequence of computing stress components in fiber and matrix. This information may be processed to predict the interfacial failure and other modes of failure.

Effective properties for each volume fraction of the representative volume element of composite are computed using the N directional stiffness and strength module of ASCA.

The constituent properties used are given below in Table 1. Table 2 gives effective ply properties with fiber volume fraction as 0.55. Table 2 also provides the effective laminate properties for (0/90/45/-45)s laminate made with these ply property.

Table 1: Constituent Properties of Fiber and Matrix Materials		
Property	AS4	H3501
E_x (msi)	34.1	0.62
E_y (msi)	2.02	0.62
υ_{xy}	0.2	0.34
υ_{xz}	0.24	0.34
G_{xy} (msi)	4.0	0.23

Table 2: Properties of a ply and of the quasi-isotropic laminate.	
AS4/3501, V_f=0.55	$(0/90/45/-45)_s$
E_{11}=19.0 msi	E_x=7.29 msi
E_{22}=E_{33}=1.23 msi	E_y=7.29 msi
v_{12}=v_{13}=.26	v_{xy}=.3
v_{23}=.38	G_{xy}=2.79 msi
G_{12}=G_{13}=.67 msi	
G_{23}=.44 msi	

The notations used in Tables 1 and 2 are conventional material properties, E, *v* and G representing Young's modulus, Poisson Ratio and Shear modulus with subscripts representing the corresponding direction. Laminate properties are calculated by using ASCA's Composite Laminate Analysis Program (CLAP) module.

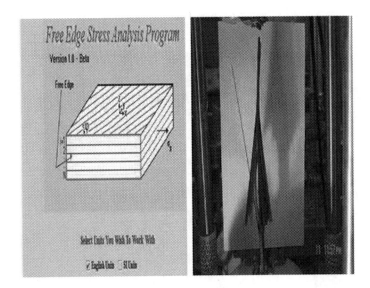

Figure 10.3: Free Edge Stress Analysis Program (FESAP) a Module of ASCA package for ply wise stress analysis and Fatigue testing of a composite laminate specimen.

Figure 10.3 shows the top cover sheet of the FESAP module of ASCA and a sample specimen in a material testing system for conducting mechanical tests.

Using the properties given in Table 2, inter-laminar stress components for applied stress in the laminate axis direction of 1 ksi were computed. These results for 0 and 90 degree ply interface are given in Figure 10.4. Inter-laminar stress components for other 3 interfaces (90/45, 45/-45 and mid surface) are given in Figures 10.5, 10.6 and 10.7, respectively. These results provide an indication as to the variation in the stress components at different locations of the laminate. Using these stress components, appropriate failure criteria can be applied to predict the failure.

10. 9 E-gls fiber and EPN 858 matrix laminates:

Using another sample composite material consisting of E-gls fiber and EPN 858 matrix, effective ply properties of E-gls/EPN 858 composite are calculated using the NDSAND module of ASCA. Five cases of different Volume Fraction (V_f) of fiber were considered for predicting the ply properties using the RVE model. These values of V_f are .55, .33, .21, .16, .13.

Table 3 provides a summary of calculations done using the Micromechanics model NDSANDS in ASCA [1]. Fiber radius F_r is .002" and matrix radius M_r has five values given in the table. Corresponding fiber volume fraction (V_f) values in the composite ply are also given. For computation of these results, we have used the following properties: Fiber material E-gls theYoung's modulus is 10.5 msi, Poisson ratio is .22 and for matrix material EPN 858 the Young's modulus is 0.42 msi, Poisson ratio is .35. These are sample materials resembling the composites to be considered. At the time of comparison with experimental values real material properties will be used.

Table 3: Effective properties of E-Gls/ EPN 858 Composite

F_r	.002"				
M_r	0.0027"	0.0035"	0.0043"	0.005"	0.0055"
V_f	0.55	0.33	0.21	0.16	0.13
E_{11}(msi)	5.8	3.6	2.5	2	1.71
E_{22}(msi)	1.39	0.84	0.67	0.6	0.57
v_{12}	0.27	0.3	0.31	0.32	0.33
v_{23}	0.35	0.42	0.45	0.45	0.45
G_{12}(msi)	0.46	0.29	0.23	0.21	0.2
G_{23}(msi)	0.52	0.3	0.23	0.21	0.2

This study is an effort to understand the effects of changing selected processing factors or characteristics in fused filament fabrication on the final material properties; the range and confidence interval for different material properties produced using specified parameters; and 3D printed aerospace parts meeting the same service specifications as traditionally manufacturing parts. As stated earlier Figure 10.1 shows the sequence of composite laminate layup. We will start with the selection of fiber and matrix materials. Based upon the above mentioned properties of fiber and matrix materials, ply properties are computed and are given in Table 3. These effective properties for each volume fraction of the representative volume element of composite are computed using the N directional stiffness and strength module of ASCA. Figure 10.2 shows schema of stress calculations for prediction of matrix and fiber interface failure.

The notations used in Table 1 are conventional material properties, E, v and G representing Young's modulus, Poisson Ratio and Shear modulus with subscripts representing the corresponding direction. Using the properties given in Table 3, stress components for applied stress in the laminate axis direction of 1 ksi, inter-laminar stress components at different ply interface are given in Table 4.

Table 4: Stress components at different locations.

y=0

Ply #	σ_z	Γ_{yz}	σ_x	σ_y
1 LS	10.25	0	149	$\sigma\Gamma\upsilon$
1 US	0	0	148	
2 LS	7.32	2068		
2 US	10.25	2068		
3 LS	5.36	148		
3 US	7.32	148		
4 LS	3.36	2067		
4 US	5.36	2067		
5 LS	-7	147		
5 US	3.36	149		

y=.005

Ply #	σ_z	Γ_{yz}	σ_x	σ_y
1 LS	-0.85	2.7	151	4.68
1 US	0	0	150	0.55
2 LS	0.7	-2.25	2060	-9.5
2 US	-0.86	2.7	2060	-11.5
3 LS	1.1	1.64	153	9.1
3 US	0.7	-2.2	153	9.2
4 LS	1.13	-1.85	2060	-8.8
4 US	1.11	1.64	2060	-8.8
5 LS	2.4	0	153	9.1
5 US	1.14	-1.86	153	10.46

y=.01

Ply #	σ_z	Γ_{yz}	σ_x	σ_y
1 LS	-0.6	1.49	152	6.6
1 US	0	0	152	4.9
2 LS	-0.205	0.369	2059	-12.07
2 US	-0.61	1.49	2059	-14.54
3 LS	0.28	0.13	153	10.21
3 US	-0.21	0.36	153	10.21
4 LS	0.57	-0.05	2060	-9.97
4 US	0.28	0.13	2060	-10.3
5 LS	0.73	0	153	10.43
5 US	0.58	-0.05	153	10.55

Stress analysis is done for the case with V_f=.55, stiffness properties given in Table 3 for hybrid laminate (m/c/m/c/m)$_s$, m represents matrix layer, c represent composite layer and m represents half matrix layer. In Table 4, LS represents lower surface of the ply, US represents upper surface of the ply. Three values of y (=0, .005 and .01) represent free edge, one ply away from the free edge and 2 plies away from the free edge respectively. These results provide an indication as to the variation in the stress components at different locations of the laminate. Using these stress components, appropriate failure criteria can be applied to predict the failure.

We intend to make laminate specimens using the 3DP process and conduct tests to measure effective response characteristics and compare with ASCA predictions. This process initially uses a 3D printed coupon. Once the model is revised for 3DP, the next phase of analysis would be to assess 3D prints of a real laminate part/component of a system such as a helicopter which was or is obsolete or hard to manufacture.

In this section, a review of 3DP technology is given, and a set of composite laminates is analyzed for predicting the response under mechanical loads. Various parameters of interest, for example: volume fraction, laminate orientation etc. are considered. The next step is to make these laminates and test for measuring their response characteristics and failure mechanisms. This section provides an iterative modeling, manufacturing, and testing program.

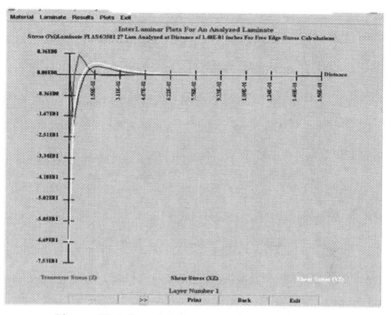

Figure 10.4: Interlaminar stress components at
0/90 interface of (0/90/45/-45)s laminate.

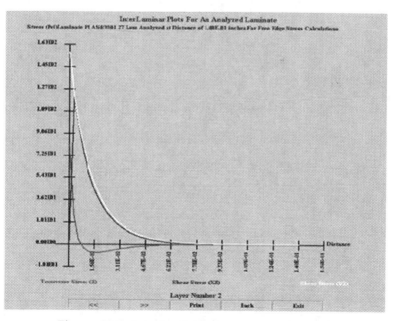

Figure 10.5: Interlaminar stress components at
90/45 interface of (0/90/45/-45)s laminate.

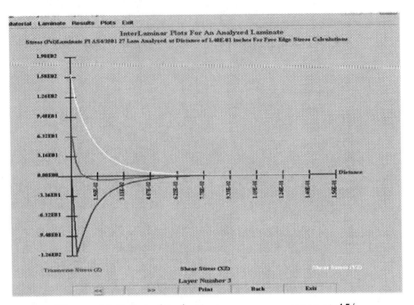

Figure 10.6: Interlaminar stress components at 45/-45 interface of (0/90/45/-45)s laminate.

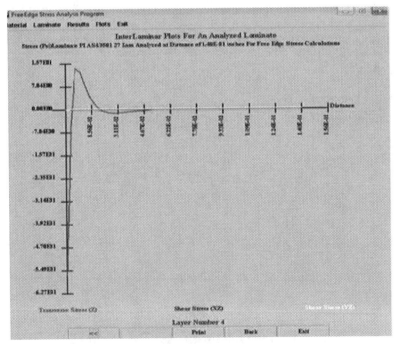

Figure 10.7: Interlaminar stress components at mid-surface of (0/90/45/-45)s laminate.

11

IMPACT OF MECH 505 LEADING TO STUDENTS RESEARCH CONTRIBUTIONS

This document provides the Journal publications during Dr. Soni's second tour of duty (Dec 11, 2005 to Dec 11, 2011) at the Air Force Institute of Technology (AFIT). These publications are results of his involvement in advising numerous MS thesis and PhD dissertation students.

Team work inculcated during these investigations has far reaching consequences. As can be seen the start takes place from a thought related to the research topic of interest to the student. That leads to continuous interaction with various faculty members, other students, computer support staff, testing laboratory staff and administrative staff. The process is rigorous and varies from student to student with relevant interests. Along the way come many challenges, frustrations and some excitements. These challenges provide excellent opportunities for students to learn from real world problems and processes of finding acceptable solutions. Publications in peer reviewed Journals are a good measure of the students' quality of work.

In these publications many challenging topics relevant to interests of the Air Force are investigated. It is hard to point out specific contributions by different individuals in the accomplishment of different research publications. It is certain that outstanding people working in AFIT and the Air Force Research Laboratory (AFRL) organizations deserve all the credit for the resulting work. It is hoped that the work continues for the benefit of all concerned past and future investigators.

Dr. Soni would like to thank Dr. Robert Calico, former dean of the School of Engineering and Management at AFIT for providing him with an opportunity to work in AFIT and make this research possible. Special thanks to sponsors whose foresight, funding and encouragement made these publications possible.

Thanks to all those who have contributed directly or indirectly, including students, sponsors, AFIT and visiting faculty; and staff.

Som R. Soni

Journal Articles

1. Jeffrey D. Kuhn and Som R. Soni, "A Design of Experiments Approach to Determining Structural Health Monitoring Sensor Durability", Tech Science Press SL, vol.1, no.1, pp.61-73, 2009
2. E.D. Swenson and S.R. Soni, "Damage Detection in a Geometrically Constrained Area", Tech Science Press SL, vol.1, no.2, pp.95-110, 2009
3. Jeffrey D. Kuhn and Som R. Soni, "Estimating Changes in SHM Performance Using Probability of Detection Degradation Functions", Tech Science Press SL, vol.2, no.1, pp.1-10, 2009
4. S.R. Soni, E.D. Swenson and R.T. Underwood, "Crack Detection in High Strain Aerospace Applications", Tech Science Press SL, vol.2, no.2, pp.63-79, 2009
5. Hitesh Kapoor, James L. Blackshire and Som R. Soni, "Damage Detection in z-Fiber Reinforced, Co-Cured Composite Pi-Joint Using Pitch-Catch Ultrasonic Analysis and Scanning Laser Vibrometry", Tech Science Press SL, vol.3, no.3, pp.221-238, 2010
6. Teresa Wu, Som Soni, Mengqi Hu, Fan Li and Adedeji Badiru, "The Application of Memetic Algorithms for Forearm Crutch Design: A Case Study", Hindawi Publishing Corporation, Mathematical Problems in Engineering, Volume 2011, Article ID 162580, 14 pages doi:10.1155/2011/162580

7. Fan Li, Teresa Wu, Adedeji Badiru, Mengqi Hu, Som Soni, "SLDM_RBDO: A Single Loop Deterministic Method for Reliability Based Design Optimization", Jl. of Engineering Optimization, 2011. Wu et al-2.pdf

8. Francisco Ospina, James L. Blackshire, Som R. Soni and David R. Jacques, "Analysis and Modeling of Small Crack Detection in Pressurized Fuselages for Structural Health Monitoring Applications", Tech Science Press SL, vol.0, no.0, pp.1-13, 2014 (in press)

9. M. Al-Romaihi, S. R. Soni, J. D. Weir, A. B. Badiru, Alfred E. Thal, and Barth Shenk, "Cost Estimating Relationships Between Part Count and Advanced Composite Aircraft Manufacturing Cost Elements", Tech Science Press SL, accepted for publication.

10. Phillip O'Connell, J. Robert Wirthlin, James Malas and Som Soni, "Application of Systems Engineering to USAF Small Business Innovative Research (SBIR)", Procedia Computer Science 16 (2013) 621 – 630, 1877-0509, Published by Elsevier B.V.

List of Thesis Advised by the author at AFIT

1. Alan Albert, Efstathios Antoniou, Stephen Leggiero, Kimberly Tooman & Roman Veglio, "A Systems Engineering Approach to Integrated Structural Health Monitoring System for Aging Aircrafts", AFIT/GSE/ENY/06-M02. 2006, http://www.dtic.mil/dtic/tr/fulltext/u2/a449297.pdf.

2. Jason Freels, "Modeling Fracture in Z-Pinned Composite Co-cured Laminates Using Smeared Properties and Cohesive Elements in DYNA3D", AFIT/GMS/ENY/06-S01, 2006. http://www.dtic.mil/dtic/tr/fulltext/u2/a456697.pdf.

3. Matthew Bond, James Rodriguez & Hieu Nguyen," A Systems Engineering Process for an Integrated Structural Health Monitoring System for Aging Aircraft II, 2007. http://www.dtic.mil/dtic/tr/fulltext/u2/a469542.pdf.

4. Jason Brown & Travis Hanson, "A Systems Engineering Process for an Integrated Structural Health Monitoring System for Aging Aircraft III, 2008, AFIT/GSE/ENV/08-M03.

5. Roman T. Underwood, "Damage Detection Analysis Using Lamb Waves in Restricted Geometry for Aerospace Applications",

2008, AFIT/GAE/ENY/08-M29, http://www.dtic.mil/dtic/tr/fulltext/u2/a482728.pdf

6 Barker, Schroeder & Gurbuz, "A Systems Engineering Process for an Integrated Structural Health Monitoring System for Aging Aircraft II, 2009, 2009-AFIT-GSE-ENV-09-M04.

7. Kuhn – PhD, "Changes in Structural Health Monitoring System Due to Environmental Factors", 2009, AFIT-DS-ENV-09-S01. http://www.dtic.mil/dtic/tr/fulltext/u2/a504996.pdf

8. Soo Chan Jee, "Shape Memory Polymer for Aircraft Structures", 2010, AFIT-GSE-ENV-10-M02. http://www.dtic.mil/cgibin/GetTRDoc?Location=U2&doc=GetTRDoc.pdf&AD=ADA516942.

9. Aaron Lemke, "Part Count: Monolithic Part Effects on Manufacturing Labor Cost etc.", 2010, AFIT-GFA-ENV-10-M02.

10. Ben Alton, "Mechanical Properties Characterization and Business Case Analysis of Metal- Polymer Hybrid Laminates, 2010, AFIT-GRD-ENV-10-M04.

11. Ed Beckett and H. Shin, "A Systems Architecture and Advanced Sensors Application for Real Time Aircraft Structural Health Monitoring" 2011, AFIT-GSE-ENV-11-M04. http://oai.dtic.mil/oai/oai?verb=getRecord&metadataPrefix=html&identifier=ADA538381

12. Phil Connell, "Systems Engineering Applications for Small Business Innovative Research (SBIR) Projects. http://www.dtic.mil/docs/citations/ADA564426

13. Dan Lambert, "Composite Aircraft Life Cycle Cost Estimating Model", 2011, AFIT-GCA-ENV-11-M02. http://www.dtic.mil/dtic/tr/fulltext/u2/a538329.pdf

14. Francisco Ospina, "An Enhanced Fuselage Ultrasound Inspection Approach for ISHM Purposes, 2012, AFIT/GSE/ENV/12-M12. http://www.dtic.mil/cgi-bin/GetTRDoc?AD=ADA557890

15. Frank Sha, "Structural Health Monitoring of a Two Layer Composite Armor System", 2012, AFIT/GEM/ENV/12-M19, http://www.dtic.mil/get-tr-doc/pdf?AD=ADA557564

12

REFERENCES

Stephen Tsai and H. T. Hahn, "Introduction to Composite Materials"

Stephen Tsai, "Theory of Composite Design"

G.H. Holister and G. Thomas, "Fiber Reinforced Materials"

Sanjay K. Mazumdar, "Composite Manufacturing- Materials, Product, and Process Engineering"

R.M. Christensen, "Mechanics of Composite Materials"

N.J Pagano," Interlaminar Response of Composite Materials", Elsevier, 1989.

Soni, Som R., Software package entitled, "Automated System for Composite Analysis", AdTech Systems Research Inc., Beavercreek, OH.

Connor, Jerome J., Faraji, Susan, "Fundamentals of Structural Engineering" Springer

Kenneth Leet, Chia-Ming Uang and Anne Gilbert, "Fundamentals of Structural Analysis 4th Edition", McGraw Hill.

http://webpages.sdsmt.edu/~kmuci/Course_Materials/ME-216/Lecture-Notes/Principal-Stresses-3D-Example.pdf

http://www.efunda.com/formulae/solid_mechanics/mat_mechanics/calc_principal_stress.cfm

http://www.engineeringtoolbox.com/stress-strain-d_950.html

http://www.differencebetween.com/difference-between-stress-and-vs-strain/

http://www.mathalino.com/reviewer/mechanics-and-strength-of-materials/stress-strain-diagram

http://en.wikipedia.org/wiki/Stress%E2%80%93strain_curve

http://www.esm.psu.edu/courses/emch13d/design/design-fund/design-notes/design_buckling/bucdeslu.pdf

http://www.ewp.rpi.edu/hartford/~ernesto/Su2012/EP/Materialsfor Students/Aiello/Roark-Ch06.pdf

http://en.wikipedia.org/wiki/Fatigue_(material)

http://ebookmarket.org/pdf/introduction-to-composite-materials-tsai-pdf

http://catalogue.nla.gov.au/Record/61254

Printed in the United States
By Bookmasters